# A Thousand Years of Whaling:
## A Faroese Common Property Regime

Seán P. Kerins

CCI PRESS
2010

Circumpolar Research Series No. 12

**Library and Archives Canada Cataloguing in Publication**

Kerins, Seán P.

A thousand years of whaling: a Faroese common property regime / Seán P. Kerins.

(Circumpolar research series, 0838-133X ;v12)
Includes bibliographical references.
ISBN 978-1-896445-52-6

1. Whaling—Faroe Islands. 2. Faroe Islands—Economic conditions. 3. Faroe Islands—Social life and customs. 4. Whaling—Environmental aspects—Faroe Islands. 5. Whales—Ecology—Faroe Islands. I. Title. II. Series: Circumpolar research series; 12.

**SH383.5.F27K47 2010** **333.95'95094915** **C2011-900233-7**



This publication was made possible with funding from Eik Visindagrunnurin (the Eik Science Foundation) of the Faroe Islands.

**Cover photo credits**: Full cover background photo courtesy Hans Eli Sivertsen. Back cover photos (from top to bottom), courtesy Bjarni Mikkelsen, Seán Kerins, Seán Kerins, Seán Kerins, Seán Kerins, Bjarni Mikkelsen, Bjarni Mikkelsen, Bjarni Mikkelsen.

Cover design by John Hughes, Australian National University.
Printed in Edmonton, Alberta, Canada by art design printing, inc.

**CCIP ISSN 0838-133X**
**CCIP ISBN-13 978-1-896445-52-6**
**Circumpolar Research Series No. 12**
**Studies in Whaling No. 9**

# A Thousand Years of Whaling: A Faroese Common Property Regime

## Table of Contents

# *Acknowledgements*


Many people in the Faroes generously gave their time and warm hospitality during my fieldwork. In particular, I am indebted to Ólavur Sjúrðarberg who encouraged me to come and live in the Faroes and learn about Faroese life, Faroese whaling and Faroese decision-making. I would also like to thank Borghild, Jonna, Tummas, Hanus and Petur Sjúrðarberg, the late Petur Emil Christiansen, Oli Høgnesen, Høgni Høgnesen, and Gunna Høgnesen who always welcomed me into their family life during my visits to Leirvík. Thanks to Absalon, Laura, Amalia, Jógvan, Jóhan, and Andrass Ebbi for their hospitality on my visits to Fugloy and introducing me to the culinary delights afforded by puffins.

I am indebted to Dorete Bloch, Director of *Føroya Náttúrugripasavn* (Faroese Museum of Natural History), for providing a roof over my head in Tórshavn. Also thanks to the staff of *Náttúrugripasavn* for their warm hospitality and assistance with the research especially Bjarni Mikkelsen (also for use of his photos), Bjørki Geyti, Astrid Andreasen, Høgni Arnbjarnarson, Lise Ofstad and Eyðfinnur Stefansson.

A special thanks to Jústines Olsen, Kate Sanderson and Kjartan Hoydal for their guidance with the research, especially Kate for her assistance with the Faroese language and providing me with much detail concerning the anti-whaling protests and the Faroese response to these activities.

I am grateful to the members of *Grindamannafelagið* (the Faroese Pilot Whalers' Association) for welcoming me to their meetings, answering my many questions and ensuring I experienced the *grindadráp* from within. Those experiences were most valuable. In this regard a special thanks to Ólavur Sjúrðarberg, Hans Jákup Hermansen, and Finnbogi Joensen.

Thanks to David Geyti, Gunner Djurhuss, and Karl Johan Jensen for their hospitality and insights into Faroese life. Thanks to Pauli Andreasen for his time in answering questions about past practices with the *grindadráp*.

I am grateful to Athol Chase and Lex Brown in Brisbane, Australia for their support and guidance during my research. Thanks also to Milton Freeman in Canada for assistance in completing this monograph, and to Elaine Maloney and Cindy Mason of CCI Press for copy editing and preparation of the manuscript for printing. Thanks to John Hughes, Australian National University for the cover design. Thanks also to the late Sir Archie Te Atawhai Taiaroa and Robin Hapi of Te Ohu Kai Moana for the encouragement—*Ka nui te mihi ki a kōrua.*

I am indebted to the late Ingmar Egede of Greenland who was a great advocate for research on community-based whaling and actively encouraged this undertaking.

This publication was made possible with funding from Eik Visindagrunnurin (the Eik Science Foundation) of the Faroe Islands.

Lastly, a special thanks to my family and friends in Aotearoa/New Zealand and Australia for their support and encouragement with this research undertaking and life's adventures.

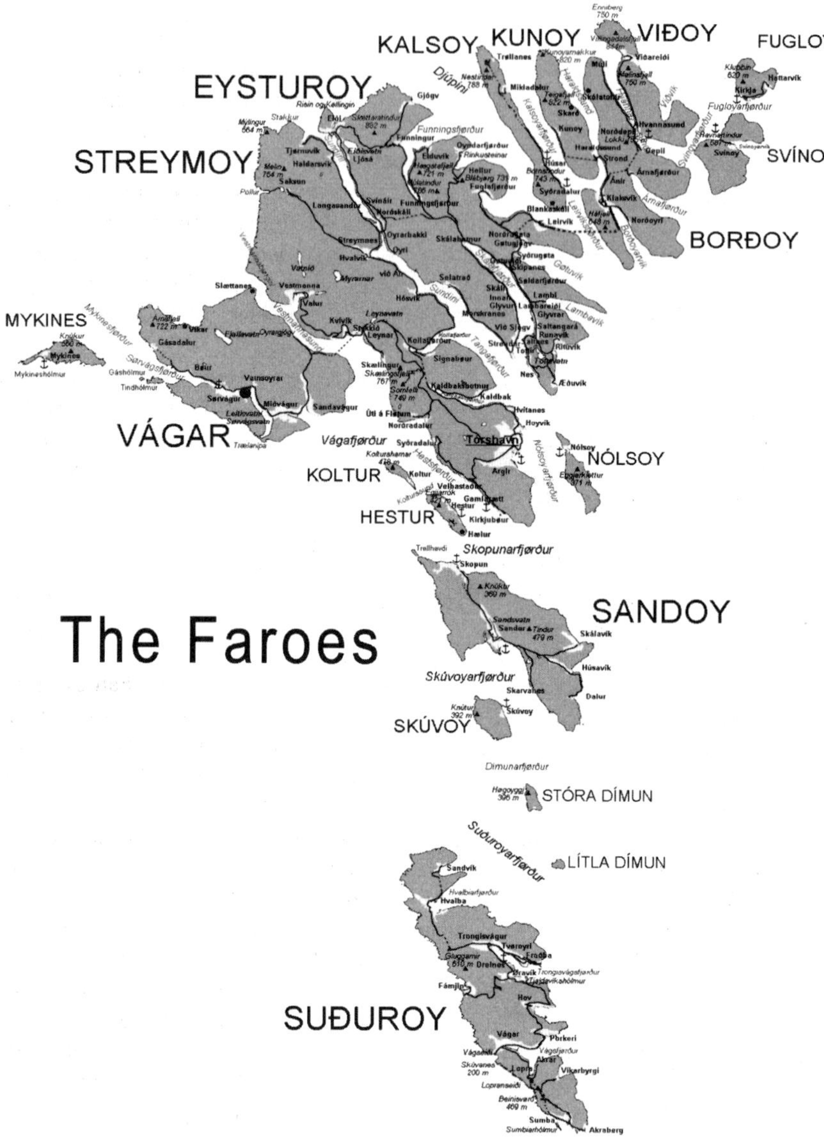


*Map of The Faroe Islands (mappery.com/map-name/Faroe-Islands-Map).*

# *Whales, Whaling and Protest: An Introduction Relevant to the Faroe Islands*

> *Pilot whaling goes to the very bone of what it means to be Faroese; it's your food, what you eat, your national diet, your culinary uniqueness. An activity that goes back for generations and generations, it [the protest action] was a focus on the very heart, the very core of what it means to be Faroese* (Kate Sanderson 2001, interview).

*Søtur er Sjálvgivin Biti* is a Faroese proverb, which roughly translates to English as: 'Sweet is the food one can provide for oneself' (Department of Foreign Affairs 2003:23). This volume relates to that particular maxim. It is about real life events and how whaling and whales feed families, communities and economies over long periods, ensuring community reliance on the products of the hunt and creating a vital nutritional, cultural and ecological link between people and whales. It is also about how a small nation on the periphery of the economic and political power centres of the world economy can, through local decision-making, shape its future in a positive manner.

The utilisation of schools of long-finned pilot whales (*Globicephala melas*), or *grind*[1] in Faroese, has been occurring continuously in the Faroes since the 9th century when it is thought early settlers began developing practices for driving the pilot whales ashore where they were slaughtered, hauled up, assessed, flensed and then divided freely among community members to be used as food.

The use of marine mammals (dolphins, whales and seals) along with seabirds, has served as an important component of the local economy of the Faroe Islands since settlement of this small North Atlantic archipelago. The use of marine mammals is important because of the islands' fragile soils, limited cultivatable acreage and the harsh climate of the region. Access to marine mammals, an abundant and nutritious food resource obtained at little cost, is especially important during times of economic hardship, *e.g.*, during the harsh trade monopoly imposed by the Danish regime from the 16th century to the mid-19th centuries, or in early 1992 when the Faroese economy collapsed and almost the entire fishing industry faced bankruptcy. At that latter time, the Gross National Product (GNP) fell by 40 percent, with unemployment rising from zero to 25 percent and migration reached 10 percent of the population (Hovgaard and Apostle 2002). During these difficult economic times, marine mammals, and the indigenous large-scale cooperative event of the *grindadráp* (the driving, slaughtering, assessment and division of pilot whales), enabled small isolated

[1] A school of pilot whales are termed a *grind* (pronounced with a short 'i' as in window) in the Faroese language. A single pilot whale is termed a *grindahvalur*.

village communities spread out over the 18 inhabited islands to secure regular supplies of highly nutritious food. This durable community-based institution, and the body of rules and processes that guide the participants in the *grindadráp,* reflect the social, political, economic and environmental influences on the Faroese people and the common-property nature of the resource itself. As the *grindadráp* became refined over the centuries, it became deeply rooted in and reflective of Faroese culture, the social, economic and political environment, and the historic struggle of the people to survive and maintain their own independence and decision-making over local resources.

In the 19th century, despite the gradual development of the capitalist market economy based on commercial fisheries, the use of marine mammals and sea birds remained important to Faroese food security. At this time, the *grindadráp* became "an established symbol of Faroese identity along with other elements of traditional Faroese life which had lost their primary economic significance with the expansion and modernisation of fishing as an industry" (Sanderson 1992:1). Others also refer to the unique place the pilot whale hunt occupies in creating a distinctive Faroese national identity (Wyle and Margolin 1981:29; Joensen 1990:182; Kalland *et al.* 2005:125).

During the 1980s, a single component of the institution of the *grindadráp*, the slaughter, was isolated from its other social, biological and economic components and portrayed negatively (using bloody images of the slaughter) through the international media by environmental and animal rights organisations as a proxy for, or representation of, the *grindadráp*. Faroese pilot whaling was portrayed as unsustainable, barbaric, and totally unnecessary in a 'modern' world, and consequently should be ended through protest action involving economic boycotts against Faroese fishery exports.

The history of industrial over-exploitation of many species of whales shaped the negative views of whaling held by urban-based people, especially in affluent industrial nations, living physically and psychologically at a distance from the animals that produce the food, fibre and hides on which they depend. The promotion of the idea that whales are a higher order than other animals, with claims of intelligence on a level comparable with humans, further polarised public opinion against all types of whaling, regardless of the abundance of the resource, the sustainability of the hunt, or its central cultural and economic role in the provision of large amounts of high quality food to coastal communities living in regions where agriculture is severely limited. Many environmental and animal rights organizations, usually based close to centres of economic and political power in industrialized nations, used these favourable circumstances for their own economic benefit and to finance further protest campaigns.

The clash of views between anti-whaling organisations and the Faroese became especially threatening in the late 1980s and the early 1990s, when numerous environmental and animal rights organisations campaigned internationally to boycott Faroese fisheries and fish products in order to undermine the Faroese economy and 'bring the Faroes to their knees.' With some 95.9 percent of the Faroese economy based on fisheries and fish products, this action clearly threatened the economic viability of the Faroes, and caused the Faroese to reflect on and re-evaluate the *grindadráp* and question its role within a modern vibrant culture and economy (*Hagstova Føroya* 2000).

This chapter introduces the reader to whales and whaling and provides a brief overview of the use of whales by humans from 5,500 years ago to the rise of unsustainable large-scale industrial whaling, a historic development which caused the collapse of many stocks of large whales. This over-exploitation, along with the need to keep whale-oil prices high, saw the world's whaling nations form the International Whaling Commission (IWC). The role of the IWC, in regard to Faroese whaling, is outlined below, along with the rise of the protest movement which initially targeted large-scale industrial whaling and then moved to seek bans on other forms of whaling. In particular, this chapter considers actions of the protest movement targeted specifically at the Faroese pilot whale hunt.

## Whales

The term 'whale' is ambiguous: sometimes it is used to refer to all cetaceans (whales, porpoises and dolphins), sometimes only to the larger of these species, and, at other times, only to members of particular families within the order *Cetacea*. This confusion stems from the common names with which some species of cetaceans have been labelled. For example, although orcas (killer whales) and pilot whales include 'whale' in their name, in a strictly biological sense these cetaceans are dolphins. In a scientific sense, whales are those cetaceans that are neither dolphins (i.e., members of the families *Delphinidae* or *Platanistoidae*) nor porpoises (*Phocoenidae*). All cetaceans are descended from land-living mammals, most likely A*rtiodactyla* (a group of herbivorous terrestrial mammals, that includes camels, cattle, deer, giraffe, goats, pigs and sheep), believed to have entered the water some 50 million years ago (Fontaine 1998).

Cetaceans, like all mammals, are warm-blooded, breathe air and feed their young milk from mammary glands. Females (cows) give birth to a single calf. Their streamlined bodies make them highly adapted to aquatic life, with the males' (bulls) genitals being retracted into a cavity. Their forelimbs are modified into flippers and the hind limbs are vestigial, unattached to the spine, and hidden entirely within the body. Their powerful tails (flukes) are horizontal and used predominantly for propulsion through the water. Their bodies are nearly hairless, with a layer of fat (blubber) directly under the skin serving as insulation and as an energy reserve. Cetaceans breathe air through blowholes located on the top of their heads, and species such as the sperm whale (*Physeter macrocephalus*) are able to remain underwater for up to two hours, diving several kilometres before resurfacing in order to breathe. The size of the various species of cetaceans varies greatly, with small river dolphins measuring from about 1.2 metres in length and weighing 40 kilograms, to the massive blue whale (*Balaenoptera musculus*) measuring up to 30 metres in length and weighing up to 180 tonnes.

The order *Cetacea* is divided into two suborders: *Mysticeti* (baleen whales) and *Odontoceti* (toothed whales, which includes dolphins and porpoises). Baleen[2] whales, the largest of all cetaceans, are filter feeders and

---

[2] Baleen (also called whalebone) is a modification of the animal's epidermis and makes up baleen plates, sieve-like structures made of

feed predominantly on planktonic invertebrates and small-bodied fish. Toothed whales feed on fish and squid, and, in the case of orcas, on seals and sometimes other whales and dolphins.

There are about 80 species of cetaceans inhabiting the majority of the world's oceans and numerous river systems. Of these species, only a small number have been subjected to organized hunts. Species hunted today include: bowhead (*Balaena mysticetus)*, Bryde's (*Balaenoptera edeni*), fin (*Balaenoptera physalus),* humpback (*Megaptera novaeangliae*); gray (*Eschrichtius robustus*), minke (*Balaenoptera acutorostrata*, *Balaenoptera bonaerensis*), sei (*Balaenoptera borealis*), and sperm whale (*Physeter macrocephalus*). These species are often referred to as the 'International Whaling Commission (IWC) Whales' as they are among the twelve species of large whales acknowledged as falling under the authority of the International Convention for the Regulation of Whaling (ICRW). Other species hunted (the so-called 'non-IWC whales') include: the beluga (*Delphinapterus leucas*), narwhal (*Monodon monoceros*), Baird's beaked whale (*Berardius bairdii*), long-finned (*Globicephala melas*) and short-finned (*Globicephala macrorhyncus*) pilot whales, pygmy killer whale (*Feresa attenuata*), false killer whale (*Pseudorca crassidens*), harbour porpoise (*Phocoena phocaena*), Atlantic white-sided dolphin (*Lagenorhynchus acutus*), and Risso's dolphin (*Grampus griseus*), as well as several other dolphins.

## Whaling

The history of whaling extends back thousands of years, with Inuit[3] whaling dating from as early as 5,500 years ago (Freeman *et al.* 1998:59). Many peoples in various regions of the world hunted whales, dolphins and porpoises for food. Often these peoples were 'indigenous,' such as the Inuit of the Arctic regions, the Makah and Nuu-Chah-Nulth of the northwestern coastal region of the U.S. and Canada respectively, the Ainu of Japan, and Maori of New Zealand. Others with long histories of community-based whaling, who do not fit the 'indigenous' label, such as the Japanese, Norwegian, Icelandic and Faroese, are termed artisanal or small-type coastal whalers[4] (Freeman 1993).

---

keratin, which are arranged in two parallel rows. They look like combs with thick hairy fringes and are attached to the upper jaws of species in the *Mysticeti* suborder. Whales use these combs for filter feeding. Depending on the species of whale, a baleen plate can be 0.5 to 3.5 metres long, and weigh up to 90kg.

[3] Inuit are indigenous peoples living in high latitudes in Russia (Chukotka), U.S. (Alaska), Canada, and Greenland, whose economies and cultures, like the Faroese, are closely related to the hunting of marine mammals (Freeman *et al.* 1998).

[4] Artisanal or small-type coastal whalers share many similarities with indigenous whalers. In artisanal whaling or small–type coastal whaling, participation in the hunt, as well as shares in the products of the hunt, are frequently shaped by kinship (*see* International Study Group on Norwegian Small-Type Whaling 1992).

The majority of whaling undertaken by indigenous, artisanal or small-type whaling communities occurred, and continues to occur only in coastal and nearshore waters. Many of these communities are in locations with climates and geography that make agriculture difficult or impossible. The commonality all these diverse communities share is that they all whaled for food (*op. cit*). Low or intermediate technologies were initially used to hunt cetaceans; vessels ranged from skin boats, dugout canoes, or small wooden outriggers to small coastal fishing boats. Methods and technologies included nets, bamboo spears, wooden-shafted harpoons with stone, metal or mussel shell-tipped detachable harpoon heads, knives and hooks, sealskin and wooden floats, and rawhide and rope lines.

Because of the relatively low-level technologies used, there was little regulation of whaling except local governance systems for resource use developed around the use of cetaceans. Given these technological limitations, hunters' need to take whales when within hunting range was offset by vessel speed, distance from the coast, weather conditions and the ability to process, store and consume the catch. During this long association with cetaceans, coastal communities came to rely on the nutritional, cultural and economic sustenance that cetaceans provided their communities. Whaling peoples' concern for their communities was linked to their concern for the health of whale populations upon which they depended (Happynook 2004).

The hunting of whales and dolphins for food continues in many regions of the world including the Arctic, the Caribbean, northern Europe, North America, Southeast Asia and the South Pacific. Specifically, it occurs (or in some cases, occurred until recently) in Canada, Dominica, the Faroes, Greenland, Indonesia, Iceland, Japan, Norway, the Philippines, Peru, Russia, the Solomon Islands, St Lucia, St Vincent and the Grenadines and the United States of America. The use of beached whales also occurs in places such as New Zealand (*op. cit*).

**The Rise of Large-scale Industrial Whaling**

The beginnings of industrial whaling can be traced to Basque whaling in the Bay of Biscay where, by medieval times, the Basque fishery had become a significant industry, with whale meat and blubber being exported to many parts of Europe. By the 13th century, the Basques had developed the technique of a mother ship carrying smaller whale boats that would pursue their target, the Atlantic northern right whale (*Eubalaena glacialis*), a slow moving, easy to kill, non-aggressive whale rich in both baleen and oil which would float when dead (Mawer 1999). Later they invented what Morton (1982:30) describes as the precursor to the modern factory ship, where on the ships' decks they secured two large iron cauldrons set in brick structures built to contain the fires used to render the blubber at sea, which was then stored in wooden barrels for transporting back to market. By the 16th century, as the Atlantic northern right whale became depleted close to the European coasts, Basque ships crossed the Atlantic Ocean to hunt the right whales in coastal waters off the Labrador coast, and Dutch whalers established shore stations in Spitsbergen to hunt the closely-related Greenland right whale (also known as the bowhead whale) at the beginning of the 17th century. British and American whalers, operating in the

Atlantic and Pacific developed whale fisheries based on Basque techniques in the 17th and 18th centuries (Tønnessen and Johnsen 1982).

The whaling industry expanded rapidly as whale oil was in huge demand, being the major source of lighting fuel before Colonel Edwin Drake discovered oil in Pennsylvania in 1859 (Friedheim 2001:30). With the invention of the steam engine, technological limitations quickly faded and industrial whaling grew in scale and extended across the world's oceans. Whaling vessels became faster and were able to pursue swifter swimming whales that had been difficult to catch in a vessel powered only by the wind or oars. In 1884, the invention by the Norwegians of the explosive grenade harpoon, fired from a cannon on a fast steam-powered catcher vessel, further added to the industry's ability to take an ever-increasing number of whale species (Tønnessen and Johnsen 1982). Small pursuit vessels delivered their catch to shore stations or mother-ships where the carcass was flensed and the blubber rendered while the pursuit vessels looked for more whales.

Unlike subsistence whaling that had been undertaken for centuries by small coastal communities,

> The modern method was almost wholly and exclusively based on the extraction of oil, which was carried out not by melting the blubber in open pans but by cooking the blubber and other highly oleaginous bone and flesh in large pressure boilers. With the introduction of the modern method the Industrial Revolution had made its entry into whaling (*ibid*: 7).

Over time, as more and more whales were caught, their carcasses were processed into a growing array of products from oil for lubricating machines, lighting street lamps, candles, soaps and ointments. Oil was also used in the tanning, textile, and confectionary industries, for making linoleum and paint, and for tempering steel. Baleen, a secondary product, was also used for making umbrellas, whips, corsets, and industrial brushes. Despite whale oil used for lubrication and lighting being squeezed out of the market by mineral oil and various other vegetable-based oils in the mid-19th century, specialist markets still existed where whale oil was highly sought after (*ibid*: 10). These specialist markets lasted well into the 20th century, where, for example, until the late 1960s whale oil was used in submarine guidance systems, leading the U.S. military to object to listing sperm whales as an endangered species (Aron *et al.* 1999:26). However, whale meat and blubber as a food was of little importance in the global marketplace.

Friedheim succinctly describes the history of industrial whaling as:

> [A] record of extraordinary rapacious behaviour. Little concern was shown for the survival of the largest mammals this world has ever known. European and American whalers from the sixteenth to the twentieth centuries hunted one species after another to the point of extinction, usually switching to smaller species when the populations of larger animals were so decimated that hunting them was no longer commercially viable (Friedheim 2001:30).

Industrial whaling was not only a threat to the various whale stocks being targeted, but also to numerous small coastal whaling communities, many of which were having their first experience of European colonisation via a rampant industrial economy which demonstrated little concern about their rights to whales, land, or other resources. The history of whaling in the South Pacific with its clash of cultures and the spread of disease and trade in muskets, is just one example (*see* Morton 1982; Tod 1982; Mawer 1999).

**Global Management of Whales and the International Whaling Commission**

Despite large-scale industrial whaling being one of the first global industries, beginning in the 16$^{th}$ century and rapidly developing with the introduction of new technologies introduced in the 18$^{th}$ and 19$^{th}$ centuries, there was no attempt to manage the fishery at a national, regional or international level for several hundred years.

In the early stages of the industry's growth there was no concept of a law of the sea. A state's governance of the sea extended to only three nautical miles, the distance a little more than the reach of a cannon ball fired from land. All resources in the ocean outside of this narrow seaward limit were deemed open access resources, belonging to no one and over which there was no governance. In the absence of governance institutions, these open access resources were rapidly depleted by a swiftly growing, technologically advanced, highly competitive, and unregulated whaling industry.

Over-harvesting became such a serious threat that the states involved in industrial whaling were forced to attempt to regulate the industry by means of an international convention drawn up in 1931. However, it was not the conservation of the world's whale stocks that was the driving force behind the development of an international agreement to regulate whaling, for "[t]he signatories tried to use an international convention not so much to solve the problem of overharvesting but the problem of overproduction" (Friedheim 2001:31).

Initially, the *International Convention on the Regulation of Whaling* (ICRW) required the signatories to take a species-based approach and restricted hunting to baleen whales. Among other regulatory measures, this convention regulated biological standards, established whaling seasons, outlawed the taking of females accompanied by calves, and established size and catch limits (quotas) for whales allowed to be taken. The convention was unique in that the collection of hunting statistics was one of the first international actions taken by states to base wildlife management on science. Although statistics were collected, there was little scientific debate about the data and sustainability of the stocks. The convention also lacked a governance institution that could monitor and enforce the rules and processes established by the states, or resolve disagreements arising from its deliberations.

Despite the intention of the ICRW to regulate whales targeted by industrial whalers, it was fatally flawed along the lines of Hardin's (1968) 'tragedy of the commons' paradigm:

> It was characteristic of early attempts to manage common property resources. It tried to regulate the taking of whales without seeking to solve the open entry problem. That is, there was little incentive for a number of parties to co-operate to not overexploit a resource if one or more of them could take what they pleased while the majority agreed to act with restraint. Those who would accept restrictions while others did not were 'suckers' (Friedheim 2001:30).

The 1931 convention soon proved inadequate for the task of managing whales, and in 1937 it was extended by an agreement signed by the major industrial whaling states. These changes saw earlier amendments to the convention supplemented by a protocol that banned the taking of humpback whales and created a whale sanctuary in the Pacific sector of Antarctic waters (*ibid*: 31).

Another major impact on global whaling occurred during the Second World War when significant portions of the world's whaling fleets were sunk; however, this did not end large-scale industrial whaling. Friedheim notes the resumption of whaling after the Second World War was anticipated in an agreement drafted in 1944 by the Whaling Committee of the International Council for the Exploration of the Sea (ICES) (*ibid*: 31). Even at this late stage in the history of whaling with declining stocks, sustainable use was still not the International Whaling Commission's (IWC) primary motivation. "In those early post-World War II days, when the whaling industry was supplying a global market for oil, the IWC was in fact an oil cartel, concerned primarily with maintaining the world price of whale oil by regulating its supply" (Freeman 2001:139).

> Support for this assertion is provided by the fact that until the 1972–73 whaling season, the quotas were determined by Blue Whale Units (BWU). The BWU expresses the allotted catch in terms of oil yield, and has no value for biological management. One BWU was set equal to 6 sei whales, 2 fin whales, or 2.5 humpbacks. The IWC's conservation record prior to the middle 1960s was characterised by its denial of the best scientific advice and its refusal to implement quotas to prevent the severe depletion of some stocks at the cost of industry profits (Aron 1988:100).

Despite the failures in the initial regime, where the short-term economic interests of the industrial whaling states dominated the IWC decision-making process, the commission was seen to be on track for developing a successful international management model for cetaceans based on sustainable use principles (Aron *et. al.* 1999). This new model was known as the New Management Procedure (NMP). However, Friedheim contends that this new model endorsed by the IWC in 1974 may have been developed too late:

> In 1972 the United Nations Conference on Environment and Development held in Stockholm demonstrated that the world community had been careless in its use of 'natural capital' and had fired the enthusiasm of a rapidly enlarging world environmental community to do better. Where better to do better than with whales and whaling? (Friedheim 2001:4).

At the 1972 United Nations Conference on Environment and Development (UNCED) a resolution calling for a 10-year global moratorium on commercial whaling was adopted (Kalland *et al.* 2005:194). It could be argued that the outcomes of the UNCED and the rise of the whale as a symbol for environmentalism in the public consciousness ensured a dramatic change in the operations of the IWC. However, an alternative argument stands just as strongly: that the major developed states that once operated industrial whaling fleets, because of falling profits and disappearing markets for whale products, no longer had an economic incentive to continue whaling. This allowed some states without a history of whaling for food purposes, to join the IWC to oppose whaling. By doing this they could demonstrate to their voting public that they had 'green' credentials while ensuring that such a move would have no social or economic implications for their own communities who had no whaling enterprises to protect.

Membership of the IWC was originally limited to 14 states and it stayed roughly that way for over a quarter century,[5] until the late 1970s when a number of states were recruited by the U.S. and anti-whaling environmental organisations in order to make effective decisions to defeat pro-whaling states in the Commission.[6] Membership of the IWC is open to any state that formally adheres to the 1946 Convention (ICRW 1946).

---

[5] Argentina, Australia, Brazil, Canada, Chile, Denmark, France, the Netherlands, New Zealand, Peru, South Africa, the U.K., the Union of Soviet Socialist Republics, and the U.S. (Ohmagari 2005).

[6] Francisco Palacio, a former Greenpeace consultant, tells how Greenpeace "[t]argeted poor nations plus some small, newly emerging independent ones like Antigua and St Lucia. They [Greenpeace] drafted the required membership documents for submission to the US State Department" and assigned themselves or their friends as scientists and commissioners to represent these nations at the whaling commission. For instance, Palacio, a Colombian citizen based in Miami, arranged to be commissioner for St. Lucia. The Commissioner for Antigua was Palacio's friend and lawyer, Richard Baron, also from Miami. Between 1978 and 1982, Palacio says, the operation added at least half a dozen new member countries to the commission's membership to achieve the three-fourths majority necessary for a moratorium on commercial whaling, which passed in 1982 (Spencer *et al.* 1991; Ohmagari 2005).

In 2006, membership of the IWC totalled 52 states.[7] Many of these states have no discernible economic interest in whaling or in whales and as Burke asserts, their unwillingness to pay their financial contributions sometimes reflects this. "However, some states that don't pay on time do have a genuine interest, mostly in artisanal whaling, but as developing nations lack the wherewithal to pay their contribution in a timely way. In 1997 over 25 percent of the IWC membership was not entitled to vote because of arrears in their contribution" (Burke 2001a:4).

Friedheim confirms this opinion:

> Because of the IWC's spotty record of dealing on time to prevent significant reductions in the stocks of some major whale species, and because major developed states no longer had much of an economic interest in sustaining the whaling industry, and because after 1972 a number of important environmental organisations (most obviously Greenpeace) that operated beyond national borders now had the domestic political influence and emotional backing of mass members in important developed states and could successfully sway governments, a drastic change in approach was made in the IWC (Friedheim 2001:5).

Such a climate resulted, in 1982, in the IWC adopting a resolution to introduce a pause in commercial whaling that was to enter into force in the 1985-86 whaling season. The 'pause' was later declared a 'moratorium,' which remains in effect.

However, the moratorium was not the end to all whaling, as it applied only to commercial whaling of the 'great whales' regulated by the ICRW. Whaling continued both inside and outside of the ICRW. Under the convention, there is a category provided for what is known as 'aboriginal subsistence whaling' under which indigenous/aboriginal peoples are permitted to continue whaling for subsistence purposes, even of the 'great whales.' Whaling also continued outside the aboriginal subsistence whale category, as the ICRW (and hence the IWC) does not regulate small cetaceans such as pilot whales, beaked whales, beluga, narwhal, Risso's dolphin and other small cetaceans. Under current IWC regulations, aboriginal subsistence whaling is permitted for the Kingdom of Denmark (Greenland, for bowhead, fin and minke whales), the Russian Federation (Chukotka, for gray and bowhead whales), St Vincent and the Grenadines (Bequia, for humpback whales) and the U.S. (Alaska and

---

[7] IWC membership in 2004 includes: Antigua and Barbuda, Argentina, Australia, Austria, Belize, Benin, Brazil, Chile, Costa Rica, Denmark, Dominica, Finland, France, Gabon, Germany, Grenada, Iceland, India, Ireland, Italy, Japan, Kenya, Mauritania, Mexico, Monaco, Mongolia, Morocco, the Netherlands, New Zealand, Nicaragua, Norway, Oman, Palau, Panama, the People's Republic of China, Peru, Portugal, the Republic of Guinea, the Republic of Korea, the Russian Federation, Saint Kitts and Nevis, Saint Lucia, Saint Vincent and The Grenadines, San Marino, Senegal, the Solomon Islands, South Africa, Spain, Sweden, Switzerland, the U.K. and the U.S.A. (IWC 2004).

Washington State, for bowhead and gray whales) (*see* Birnie 1985; Doubleday 1989; Gambell 1993, 1997; Burke 2001b).

**Management of Small Cetaceans**

The debate about who has the authority to manage small cetaceans has been ongoing since the early 1970s when the IWC scientific committee[8] began examining the stocks of minke and other small whales. Within the debate about who has authority to manage small cetaceans there are two opposing views. On one side are those, including Australia, New Zealand, and the UK, who assert that small cetaceans should be managed within the IWC, and on the other side, those, such as the Faroes, Greenland, Japan, and several small Caribbean nations who assert that the IWC has no authority to manage small cetaceans. Despite this debate continuing for over 30 years, agreement has still not been reached (Andresen 1989; Ohmagari 2005).

The reason for this is that the gulf between these divergent views has been growing. Countries that sustainably utilise small cetaceans for food, such as Greenland and the Faroes, do not want to surrender their long-established common property rights to these species to the IWC, who in their view will usurp local decision-making over the provision of locally produced food, and place this decision in a highly politicised global institution far removed from the communities who utilise whales for food.

The Faroes has been providing information on pilot whaling to members of the IWC for many years, through its representation within the Kingdom of Denmark delegation, despite the fact that the IWC is not recognised by the Faroese as a competent body to manage small whales. The Faroese Government also exchanges information on a bilateral basis with other governments wishing to be kept updated on whaling activities in the Faroes.

Although attending some IWC meetings, the Faroese Government is not a contracting party to the ICRW. Politically, as discussed below, the Faroes form an autonomous entity within the Kingdom of Denmark; however, as an original signatory to the ICRW, the entire Danish Kingdom is represented by Denmark's membership. While Denmark itself does not whale, whaling occurs in both

[8] There has been much conflict between the IWC and its Scientific Committee (*see* Heazle 2006), a conflict that became serious in 1993 with the resignation of the Chair of the IWC Scientific Committee who wrote: "...what is the point of having a Scientific Committee if its unanimous recommendations on a matter of primary importance are treated with such contempt? And in what position does this leave the Chairman?

I have come to the conclusion that I can no longer justify to myself being the organiser of and spokesman for a Committee whose work is held in such disregard by the body to which it is responsible. Nor can I justify asking other members of the Committee to spend their valuable time working hard during the year and even harder at annual meetings knowing how the results of this work may be treated. And I cannot justify to my employer spending a significant amount of my own time to such an unproductive end.

I am left with no alternative, therefore, but to resign as Chairman of the Scientific Committee" (Hammond 1993).

Home Rule jurisdictions of the Faroes and Greenland. Therefore, the Danish position within the IWC is decided through consultations between Denmark and the Home Rule governments of the Faroes and Greenland.

Agitation by environmental and animal rights' organisations influenced the deviation from traditional IWC policy when the IWC, at its 1985 annual meeting, tasked its Humane Killing Working Group (HKWG) to study methods employed in the Faroese pilot whale drive fishery and report the results at the 1986 annual meeting. The Danish and Faroese authorities agreed with this proposal, regarding it as an exception from the general rule, with a view to clarifying an issue that, at the time, caused concern among some delegations to the IWC. The Faroese authorities produced relevant and factual information on the matter for the HKWG discussions (Olafsson 1988).

In light of the HKWG report, the IWC adopted a resolution on Faroese pilot whaling, expressing concern about the methods used and the organisation of the hunt. Recognising that some progress had already been made in the measures taken by the Faroese Government, the IWC urged the Faroes to make every effort to further improve the killing methods. The Danish Government undertook to report to the Commission on progress in relation to this matter (*ibid*: 11).

At the 1987 IWC meeting, the issue was again briefly discussed, and impatience was expressed by some member states with respect to further reporting on the issue. Despite its reservation regarding the legal competence of the IWC to engage in matters pertaining to small cetacean (i.e., 'non-IWC whale species') matters, Denmark did not want to appear to be avoiding a contentious issue, and undertook to report back in 1988, as requested at the 1986 IWC meeting (*op. cit*).

In the 1988 IWC meeting, Denmark stressed that the report was submitted for information only, and was limited to the specific issues raised by the IWC in 1986. Further material was available, but for reasons of principle would not be tabled at the IWC but would be supplied to interested parties upon specific request. The IWC took note of the report and did not put the pilot whale issue on the agenda of the 1989 IWC meeting, respecting the view that pilot whaling in the Faroes should not be on the agenda as the only small cetacean issue (*op. cit*).

Until recently, Faroese representatives attended the IWC as part of the Danish delegation; however this participation has now ceased. "As far as our participation in the IWC is concerned, the attitude of the Faroese and Danish authorities has been based upon the fact that the IWC does not regulate catches of small cetaceans, such as pilot whales" (Sanderson 2006, personal communication). The North Atlantic Marine Mammal Commission (NAMMCO), in which the Faroes is a full member, attends the IWC as an observer, and through Faroese participation in NAMMCO, as well of that of Denmark in IWC meetings, the Faroese Government is fully informed of IWC decisions regarding 'the IWC' whales.

## Protests Against Whaling

With the rapid collapse of many whale stocks across the globe making large-scale industrial whaling uneconomic, it is not surprising that 1982 saw states no longer reliant on whales as a source of oil or food form an anti-whaling coalition within the IWC. The coalition led by major ex-whaling states (including Australia, Netherlands, New Zealand, the U.K. and the U.S.) none of whom—with the exception of the U.S.—whaled for subsistence, were instrumental in promoting the resolution that, in 1982, established a pause in commercial whaling.

However, these several states, along with others who no longer relied on the products of whaling for food, economic or cultural purposes, quickly

> [r]ecognised that it was no longer possible to defend a complete ban on the grounds that conservation rules imposed by the earlier NMP [science-based 'New Management Procedure'] (and now updated in the RMP [Revised Management Procedure]) were not working or capable of working well, and they shifted their rationale [for a whaling ban] to moral and ethical grounds inside as well as outside the IWC (Friedheim 2001:5-6).

This change in position by some IWC member states was relatively easy to achieve, as will be discussed further below, because of differing modes of production of societies and how this has influenced their perception of animals and their use by humans.

The end of large-scale industrial whaling facilitated what some have described as a change in the motives behind the environmental and animal rights' organisations' operations. Hoel asserts that the number of 'environmental' organisations dependent upon public donations, soon realised the huge fundraising appeal that marine mammals offered their organisations, given that communities continuing marine mammal hunts were small, often isolated and politically weak in comparison with other adversaries, such as the highly organised and politically influential industrial polluters (Hoel 1990). Others concur with this conclusion, noting: "that while the public-awareness campaigns about the methods and behaviour of the whaling industry are laudable they are also simplistic and misleading—often deliberately so, to attract funding and support" (Aron *et al.* 1999:28).

An example of this type of fundraising is evidenced by a Greenpeace campaign, carried out in Austria in October 1992, which placed a large obituary in a major newspapers saying:

> It is with great sorrow that we announce that our friend and brother the MINKE WHALE, has passed away after being brutally murdered. Instead of wreaths and flowers, please send money to the environmental organisation Greenpeace's Whale Rights Campaign (Sørensen 1994, at: www.highnorth.no/Library/Movements/General/th-en-mo.htm ).

Commenting on the fundraising potential of marine mammals in relation to the early years of the 'Save the Seal' campaign, a Greenlander, Finn Lynge writes:

> ...Inuit income fell to an all-time low of about $100 [per hunter]. However, the anti-seal hunt organisations were strengthened financially to a remarkable degree. In the middle of the 1980s, IFAW [International Fund for Animal Welfare] made more than $6 million on its seal campaign. In the United States alone, the seal pups reaped $5 million for Greenpeace. Defenders of Wildlife (USA) got over $1.7 million and in 1981 the Centre for Environmental Education (also USA) which specialised in the anti-seal hunt campaign, made $2.2 million... In 1982 IFAW used $1 million dollars for a house with offices and computer equipment, as well as a helicopter and airplane. In 1984 IFAW threatened to get involved in a Canadian parliamentary election by means of a $3 million political campaign aimed at candidates that supported the hunt (Lynge 1992, cited in Freeman *et al.* 1998:159).

Similar levels of political support for anti-hunting campaigns continued into the 1990s, for example, "[t]he IFAW, through an affiliate organisation headed by five IFAW staff members, gave about $2 million to the British Labour Party for the 1997 election campaign" (*op. cit*).

Substantial funds such as these and the people and resources they can deploy are powerful forces working against small hunting communities in remote locations who have limited resources available to offer counter information to the wider public on their cultural and economic reliance on marine mammals (for an extensive discussion of this issue, *see* Wenzel 1991; Kalland *et al.* 2005).

**The Human–Animal Relationship**

As discussed earlier, anti-whaling member states of the IWC who recognised that it was no longer possible to defend a complete ban on whaling on the grounds that conservation rules imposed in the Revised Management Procedure (RMP) were not working or capable of working well, shifted their rationale for opposing whaling from one based on the principles of sustainable use to one based on moral and ethical grounds. This change in position by anti-whaling IWC member states was relatively easy to achieve because of the changing public perception in the developed world, both of animals and the newly emerging concern over the human–animal relationship.

This new perspective on the human–animal relationship has been shaped in part, by the capitalist mode of production that has, over the last hundred years, significantly influenced where and how people live. In particular, the growth of the manufacturing industries, and their huge demand for workers, caused a massive migration from rural areas where people were employed on farms, to the growing urban centres where manufacturing and supporting service industries were located, as dramatically illustrated by Bulliet's (2005:3) observation that "in 1900, some 40 percent of the U.S. population lived on farms. By 1990, the proportion had fallen to 2 percent." Such a marked change in the way people live has brought about "the dawning of a new era in human-animal relations" (Bulliet 2005:2).

Bulliet, in his discussion of this new era of human–animal relations, proposed two new key concepts to describe societies' relations with animals: 'domesticity' and 'postdomesticity.' An understanding of these two concepts, and differing modes of production which give rise to their divergent systems of ideas, provides an understanding of the reasons why some member-states were able to shift their rationale for opposing whaling from one based on the principles of sustainable use to one based on moral and ethical grounds:

> Domesticity refers to the social, economic, and intellectual characteristics of communities in which most members consider daily contact with domestic animals (other than pets) a normal condition of life: in short, the farming existence of a bygone generation of Americans [and others such as British and Australians], but contemporary reality for most of the developing world. Postdomesticity is defined by two characteristics. First, postdomestic people live far away, both physically and psychologically, from the animals that produce the food, fibre and hides they depend on, and they never witness the births, sexual congress, and slaughter of these animals. Yet they maintain very close relationships with companion animals—pets, often relating to them as if they were human. Second, a postdomestic society emerging from domestic antecedents continues to consume animal products in abundance, but psychologically, its members experience feelings of guilt, shame and disgust when they think (as seldom as possible) about the industrial processes by which domestic animals are rendered into products and about how those products come to the market.
>
> Domestic societies take for granted the killing of animals and experience few moral qualms in consuming animal products. By contrast, postdomestic societies, which include a steadily increasing portion of the U.S population, an even greater portion of the British population, and significant numbers in Australia and (less so) Europe, are fully immersed in emotional contradictions inherent in postdomesticity. Meat, leather, and test animals are hard to give up, but details about what goes on behind the scenes to provide these goods and cultural services are revolting. Pets and animals evoke positive feelings, but domestic animals feeding the consumer market are morally troubling (Bulliet 2005:3).

The majority of IWC member-states who shifted their position on whaling from a sustainable use approach to one based on moral and ethical grounds have the majority of their populations living in what Bulliet describes as postdomestic societies (Bulliet 2005:2-3). These postdomestic societies are in sharp contrast to those who live in what Bulliet refers to as domestic societies, such as the Faroese whose mode of production to secure and process marine mammals means that they are very aware of the bloody reality of utilising animals for food.

Hermanson described the domestic nature of Faroese society:

> We live in a place [...] with highly educated people, but these people are brought up in an area where you participate in everyday life, which might mean killing a sheep, [going] out fishing, or grow their own potatoes [...] and it is very different from life in the cities outside the Faroe Islands. Therefore, young people are interested in participating. They are interested in having an alternative; being part of everyday life and bringing this tradition forward (Hermanson, interview 2002).

Freeman (1997) suggests that this human–animal relationship goes deeper than just the domestic–postdomestic paradigm in that there is a cultural basis for some people opposing the hunting of marine mammals. This cultural opposition, he says, is rooted in western cultures' Christian belief system, where the Garden of Eden was a vegetarian place for both human and animal occupants. Following the biblical flood, God allowed humans to eat animal flesh, provided it was bloodless and was taken from herbivorous animals. This practice of bleeding slaughtered animals remains today:

> The cultural importance in European-based societies of vegetalizing meat (to render it more acceptable) is accompanied by various symbolic and semantic actions. For example, in the butchering process, the carcass is 'flowered,' by making cuts that resemble plant fronds (known as 'palm leaves' in the trade), or by displaying the raw meat in butcher shops surrounded by parsley or other greenery (*ibid*: 9).

The trend to disassociate blood from meat continues. Today, for example, when buying meat in supermarkets in developed countries, the pre-packaged portion of meat sits on top of a small absorbent pad that ensures no residual blood remains visible in association with the meat.

In western countries, these methods of presentation of meat are in sharp contrast to those occurring on the Faroes, where the slaughter and processing of animals is on public view and the person who does the slaughtering also carries out the butchering. Today, in industrial nations, as Freeman notes, the slaughter and processing of farm animals are usually separate and distinct processes, with the slaughterer and the butcher being two distinct people often working at considerable distance from each other in separate places of work (*op. cit*). With meat readily available to urban people packed in small plastic packs, the process of killing is hidden. So much so, that:

> ...nowadays slaughtering has become an invisible, exiled, almost clandestine activity. We know it goes on, of course, but it is an abstract kind of knowledge. We do not wish to eat corpses (we are carnivores, not carrion-eaters), so animals have to be slaughtered. But

> we demand an ellipsis between animal and meat (Vialles 1994:5 cited in Freeman 1997:9).

**The Whale as the Symbol of Environmental Integrity**

As discussed earlier, the term whale is an ambiguous term; sometimes it is used to refer to all cetaceans, sometimes just to the larger cetaceans, and at other times, to members of particular families within the order *Cetacea*. Sometimes the word's ambiguous nature is promoted deliberately to influence perceptions for political reasons. For example, one often hears the slogan, "Save the Whale" (*see* Fig. 1). But what does this slogan actually mean? If someone used the slogan, "Save the Bird," we would probably ask, "What bird?" Why does this questioning rarely occur when it comes to the slogan "Save the Whale?"

**Figure 1**. *A Greenpeace newspaper advertisement: 'Saving the whale from extinction': But what species of whale?* (*The Australian,* undated).

The answer to this question lies partially in the large media campaigns run by anti-whaling organisations over the last 30 years, in which these organizations often refer to 'the whale' as a single entity. Reference to 'whale' as a single entity occurs so often that it has led to the creation of what Kalland has termed the 'super whale':

> ...we are told that the whale is the largest animal on earth (this applies to the blue whale), that the whale has the largest brain on earth (sperm whale), that the whale has the largest brain in relation to body weight ratio (the bottlenose dolphin), that the whale sings pleasant and varied songs (humpback whale), that the whale is friendly (the gray whale), that the whale is endangered (the bowhead and blue whales), and so on. By talking about *the* whale, an image of a single whale possessing all of these traits emerges. But such a creature does not exist. It is a mythic creation—a 'super-whale,' which has come to represent all species of cetaceans (Kalland 1993:126).

In this manner, the 'super-whale' embodies the traits of many different species of cetaceans to attract public attention, even though it does not exist in reality. However, through the work of a growing number of environmental organisations in the 1970s this 'whale' became a new symbol representing environmental integrity. Moreover, as Sidney Holt (1985:12) argues: "Saving the whale is for millions of people a crucial test of their political ability to halt environmental destruction."

Friedheim states that the promotion of the mythical super whale creates problems for managers of these resources. He says:

> A distinctive feature of managing whales—and elephants—is that many people in developed states think of them as 'special creatures' because of their size or purported intelligence and therefore assert that they require a special regime not necessarily congruent with regimes for other forms of wildlife. Often, human-like attributes such as the ability to 'talk' and 'sing,' compose poetry, and have complex thoughts, the quality of being majestic or—if they are porpoises or orcas—of being 'cute,' are assigned to animals that are now being characterized as 'charismatic megafauna' (Friedheim 2001:25).

Sørensen warns against attributing all these characteristics to the super whale, as the super whale emerges as almost human, and sometimes even surpasses humans. "When we claim equality between human beings and whales, killing whales becomes murder, and the whalers become murderers... .While the whalers are depicted as murderers, the opponents of whaling characterise themselves as agents of what is good" (Sørensen 1994:28).

Such views of the 'super whale' do make their way into the policy arena as evidenced in the *United Nations Conference on Environment and Development Prep Com III* in Geneva in 1992, where a statement issued by Ian

Stewart (at that time, the New Zealand Commissioner to the IWC) through an intervention, listed ethical and moral considerations as the main justification for New Zealand opposing the sustainable use of cetaceans (Stewart 1991). Stewart asserted whales should not be included in the discussions on the sustainable use of living marine resources because they "have brains larger than and almost as complex as our own, making them in a sense the equivalent in the marine environment of human beings in the land environment" (EBCD 1992:3).

## Protests Against Faroese Whaling

Despite whaling continuing in many other regions of the world, and particularly in indigenous communities, the environmental and animal rights organisations did not target indigenous communities as part of their anti-whaling campaigns.

Part of the reason lies in the stereotyped views held by people in postdomestic societies about indigenous peoples. Indigenous whaling peoples—often Aboriginal and living in isolated communities in the Arctic, and with no agricultural practices possible—were seen to have limited opportunities to supplement local food (including whale meat and blubber) obtained through hunting with acceptable or nutritionally-equivalent imported alternatives. It might not have looked politically correct to target people who were already widely stereotyped as being vulnerable. In addition, such a campaign against 'vulnerable people' might not have been very successful in raising funds. A softer, non-Aboriginal target would have to be found. For a number of reasons the Faroese made an ideal target for these organisations to continue their anti-whaling campaigns and fundraising activities.

First, the Faroese were not considered an 'indigenous' people, despite being the first permanent settlers to the islands, speaking a distinct language, practising a unique culture, being colonised by Denmark, and only achieving limited Home Rule Government in 1948.

Second, being considered merely provincial citizens of Denmark, they were seen as having access to the global market economy and all the imported goods it offered, including food.

Third, they whaled outside the jurisdiction of the ICRW.

Fourth, they hunted without any quota limitations. Faroese pilot whaling is one of the largest whale hunts of a single whale species anywhere in the world, which peaked during the mid-1980s when the take averaged 2,063 per annum (Desportes *et al.* 1994).

Fifth, their whaling was not selective; Faroese whalers take all the members of a pod in a hunt, from the dominant bulls to the lactating females and their calves.

Sixth, whaling is not a process hidden from public view. It is a large public event with the slaughtering taking place on the shore in front of the whole community and any visitors who may be present. It is a process where children watch and touch the animals after slaughter. It is done in full view of cameras, enabling those watching to capture the images of the bloody slaughter of dozens of animals. Such images are particularly shocking to those living in postdomestic societies where the slaughter of animals is hidden from public view.

Faroese pilot whaling came to the attention of the anti-whaling environmental and animal rights' organisations in 1985, at the same time as the introduction of the IWC moratorium. The hunting of pilot whales by the Faroese became a *cause célèbre* for a wide variety of these organisations mostly based in the postdomestic societies of the U.K. and the U.S. Organisations involved in the protest against Faroese whaling included:

- Animal Rights Kollective—ARK II;
- Canadian Animal Rights Network;
- Environmental Investigation Agency (EIA);
- Friends of the Earth;
- Fund for Animals;
- Greenpeace;
- International Fund For Animal Welfare (IFAW);
- Marine Protection Group UK
- Peoples Trust for Endangered Species;
- Royal Society for the Prevention of Cruelty to Animals (RSPCA);
- Sea Shepherd Conservation Society;
- Whale and Dolphin Conservation Society (WDCS);
- Whale Protection Fund, and
- World Society for the Protection of Animals (WSPA).

The protest actions undertaken by these organisations took five forms, as discussed below.

**Public Advertising Campaign**

The first campaign focussed on public advertising. This involved the production of a large number of newspaper advertisements and stories, booklets, postcards, fliers for letterboxes and advertisements in cinemas. They all shared eye-catching and emotion-provoking titles such as "Hacked to Death," the "Bloody Times," the "Global War Against Small Cetaceans," and images of the slaughter and the sea turning red from the whales' blood (Fig. 2). This media material informed the public that the 'endangered pilot whale' was being 'slaughtered for fun' or the hunt was a 'blood-sport,' or the meat was not being eaten and the 'carcasses left to rot' on the beaches. Other material clearly demonstrated postdomestic views of animals, where the involvement of children was strongly condemned. For example, the Marine Protection Group UK claimed in one of its newsletters that "the hunt has an unpleasant ritualised aspect to it and the involvement of children in the slaughter serves only to add to the macabre nature of the event. It can only slow down any hope of a new attitude to these animals occurring among the islanders" (Marine Protection Group UK, undated:1).

All of the material being presented served as disturbing information for those living in postdomestic societies. One of the primary aims of this form of campaign material was to shock the postdomestic public into doing something, ideally sending money to the protest organisation, or, if not, write letters condemning the hunt.

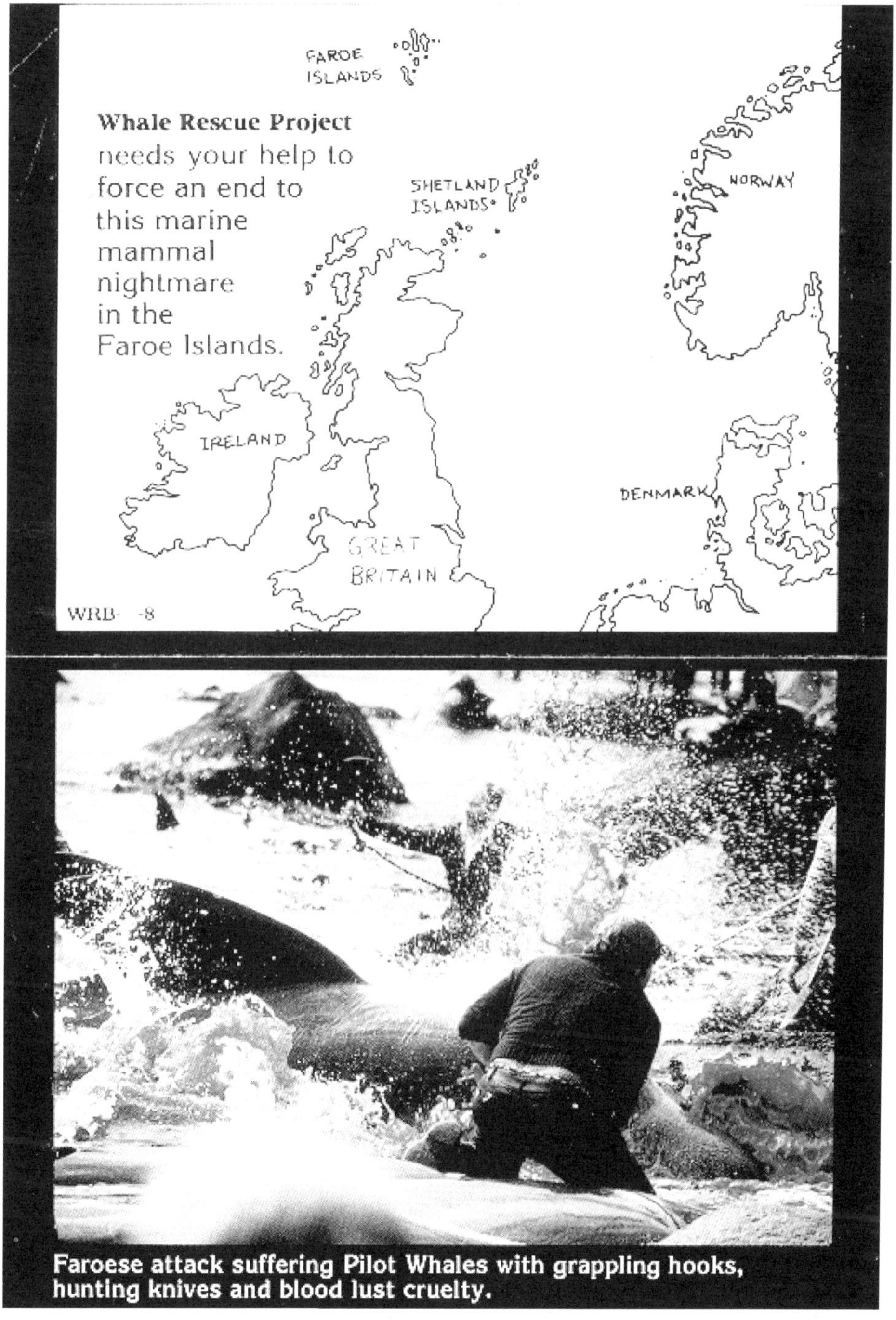


**Figure 2.** *An example of an anti-whaling campaign leaflet sent by animal rights organisations to the 'postdomestic' public* (Whale Rescue Project, undated).

**The Letter Writing Campaign**

A large and highly organised letter writing campaign, an important component of the protest, urged the public, predominantly in the U.K. and the U.S, to either write an individual letter or sign a form-letter or post-card, and mail it to the Faroese Prime Minister or the Danish Prime Minister telling them how unhappy the writer was with Faroese whaling, and calling on the whaling to be immediately stopped. During this campaign, in the late 1980s and early 1990s, the Faroese Government estimated it had received thousands of protest letters and cards which at the height of the campaign reached "a level of 1,000 missives per week" (Nauerby 1996:157).

The following is an example of one of the more extreme, but not uncommon, letters sent to the Faroese Government during the campaigns.

> From England—I have just seen on television your murderous killing of whales. I and my workmates agree that your [sic] just shit. I am a grown man and I cried to see your inhuman [sic] acts. I would love to stick one of those hooks in your head and hear you scream the way the whales did. I hope you and your family die by drowning and when you meet god [sic] he sends you back as a pilot whale so you can feel the pain the whale does and die in agony. If the atomic bomb is dropped I hope it is on the Faroe Islands, the rest of the civilised world won't miss you. The worst of health may you die soon, (signed) A normally quiet non-violent Englishman (Anonymous 1989, cited in Sanderson, 1992:104).

**The Sabotage Campaign**

This form of protest action was more direct, as witnessed in Vestmanna in July 1985 and Gøtu in July 1996, where efforts were made to sabotage pilot whale hunts in an attempt to drive the whales out to sea before they could be beached. This type of protest action was limited to a very small number of anti-whaling groups, and appears to be more about getting greater publicity and fund-raising for their organisations rather than bringing a halt to Faroese whaling.

**The Economic Boycott**

The most threatening type of protest action occurred when some of the environmental and animal rights' organisations worked together in an attempt to organise a boycott of Faroese fish products sold in European markets.

In March 1993, the so-called "Pilot Whale Campaign" was launched in London at the Queen Elizabeth II Conference Centre. This campaign was a joint project by three animal rights groups, namely the EIA, WDCS, and WSPA. Here, EIA director Alan Thornton stated the aim of the campaign was "to bring the Faroes to their knees" through consumer boycotts of Faroese fish products. Faroese representatives who tried to attend the launch to explain the Faroese case for whaling were removed from the event (Faroe Islands Solidarity Committee 1996). This campaign was of particular concern to the Faroese as about 96 percent of their total exports are based on fish and fish products, and any downturn in sales would have a significant impact on the economy and future growth of the small island nation (*Hagstova Føroya* 2000).

The campaign targeted overseas supermarkets that sold Faroese fish products. Supermarkets were threatened with boycotts if they did not stop stocking Faroese fish products; those that declined had protesters picketing outside their stores to force them to conform. Some supermarkets that refused to stop selling Faroese fish did agree to label the fish products with a country of origin sticker so that consumers could exercise a choice. During this campaign, the WDCS, in a full-page newspaper advertisement in *The Times* on May 1 1996, claimed:

> Intense pressure works, public opinion has now forced major supermarket chains including Marks and Spencer, Tesco, [and] Sainsbury to stop selling Faroese fish or to agree to label their fish so people can make their own choices. So far we believe the campaign has cost the Faroese economy over £12 million in lost and unplaced fish orders (*The Times* 1996).

**The IWC Campaign**

The fifth form of protest was an attempt to remove Faroese decision-making for whales within the oceans surrounding the Faroes, something that the Faroese have held for over one thousand years, and place that decision-making within the IWC. Here, environmental and animal rights organisations lobbied IWC member states to have the ICRW extended to include small cetaceans. This protest action saw states such as Brazil "contemplate [the IWC] progressing from scientific advice [for small cetaceans] to jointly agreed management measures for implementation by coastal States…"[9] (IWC 1993).

**Environmental and Animal Rights Organisations**

Because many of the anti-whaling environmental and animal rights organisations promote similar views, much of the campaign material is repetitive. For this reason, the nature and activities of only a selected few organisations are provided below; however, those selected are representative of organizations involved in these campaigns.

**The Environmental Investigation Agency**

The Environmental Investigation Agency (EIA) is a London-based animal rights group established in 1984. One of its directors, Alan Thornton, was formerly active in Greenpeace, establishing Greenpeace offices in London in the 1970s (Sanderson 1990). EIA began its campaign against Faroese pilot whalers in 1984 after two independent researchers spent a short time on the Faroes investigating the pilot whale hunt. Their aim was to determine if it would make a good target for an EIA fundraising campaign (Sanderson 2006, personal communication).

---

[9] Such an action would effectively have seen the loss of local decision-making by small coastal communities who hunt small cetaceans, as the IWC could have brought in zero quotas for small cetaceans.

Initially EIA took photographs and video footage of the pilot whale hunt. They used these images in campaign material, *e.g.,* in their booklet entitled *The Global War Against Small Cetaceans* (1991). They were also active within the IWC as an NGO lobbying for the halt to pilot whaling (EIA 1991, 1994).

In a report published in 1996, the EIA called for an end to the pilot whale hunt because:

> The large number of pilot whales killed casts even greater uncertainty on the population of cetaceans about which very little is known. The status, abundance and stock identity of the long-finned pilot whales in the North-East Atlantic all remain open to question… . The Faroese Government ignores the lack of information on the status of pilot whale populations exploited by the Faroe Islanders, the threats to their environment and a hunting method banned throughout all of Europe (EIA 1996:16).

The *International Field-sports and Conservation* magazine, linked to the EIA, also called for an end to the pilot whale hunt, claiming that: "Since 1979, use of modern technology had tripled the annual number of whales killed from just over 800 to over 2,400" (*International Field-Sports and Conservation* 1991:46). Another of EIA's directors, Jennifer Gibson-Lonsdale, claimed:

> There can be no justification for killing Pilot Whales in the Faroes today when there is no economic need for the meat: there is no evidence that the population can sustain the rising numbers killed: the method of killing is unacceptably cruel: [and] it is unsafe to consume the meat and blubber (Gibson-Lonsdale 1990:51-52).

The EIA website in 2006 clearly states EIA's objectives in campaigning against whaling. It states:

> Over the past 25 years, the battle to save the whales and dolphins has symbolised the dawning of a more environmentally conscious age with the determination to fight against those intent on destroying our natural world for short-term profit. EIA's aim is to ensure the safety and survival of all the world's remaining whales, dolphins and porpoises (EIA 2006).

On May 25 1995, a program was made by EIA called 'Animal Detectives.' This was devoted to Faroese pilot whaling, and was broadcast on the ITV network in the U.K. The EIA's newsletter, *News Spring,* described the program in the following way:

> EIA Investigators in your living room? A new TV series will soon beam the work of the EIA directly to millions of ITV viewers nationwide. Called 'Animal Detectives,' the series follows EIA

> undercover investigators as they expose the threats faced by the world's wildlife… (EIA 1995:2)

The film was introduced with the dramatic claim that viewers would see 'undercover agents' from the EIA risk their lives and expose a world-wide trade in wild animals. The statement bears no relationship to either the content of the film or the actual experiences of the 'undercover agents' in the Faroes.

The Faroese saw the program as being "characterised by partial, misleading and erroneous information; all seemingly tailored to reinforce EIA campaign themes" (Hoydal 1995:2-3). The Faroese Government lodged an official complaint with the U.K. Independent Television Commission (ITC) on November 1, 1995. The letter signed by Hoydal, the Faroese Director of Fisheries, stated:

> Our principal complaint is that the film did not meet the requirements for due impartiality as stipulated by the *Broadcasting Act* 1990 and in particular that Carlton TV failed in their obligation 'to ensure that justice is done to a full range of significant views and perspectives' (ITC Programme Code para 3.4) (*ibid*: 1).

Hoydal casts light on one key event that was filmed and places it within a context:

> When footage of a pilot whale hunt was first shown, it was described as being 'typical.' In fact, the film was of a badly conducted hunt, which took place in Funningsfjøður in 1992. The breaches of the pilot whaling regulations which occurred on the occasion led to the Faroese Government revoking Funningsfjøður's status as an authorised whaling bay—it was only restored when the necessary improvements had been effected. Although the footage of the Funningsfjøður hunt was used extensively, there was no acknowledgement that the events which the EIA filmed taking place that day had been deemed totally unacceptable by the Faroese authorities as well as by the public generally (*ibid*: 5).

Hoydal's letter also asserted:

> As can be seen from the attached copy of extracts from their Newsletters, the EIA had a proprietorial attitude to the series and a keen regard for the fundraising and promotional opportunities it offered. They saw it as 'our' series which was to form part of a campaign strategy. In relation to the Faroe Islands, the campaign with which it could mesh was the boycott of Faroese fish products which the EIA have been spearheading for over two years. (The boycott is aimed at forcing an end to the traditional pilot whale hunt.) In this role the EIA have become a party to a dispute with the Government and People of the Faroe Islands. The conflict reflects different philosophies on sustainable use of wildlife and embodies a

> clash of values between urban based animal rights activists and communities living close to nature.
>
> The EIA is a limited company for whom the boycott campaign is a commercial undertaking. It provides a valuable source of income by way of both donations in response to fundraising around the issue and receipts from the sale of photographs and film footage. The economic health of the company and the livelihoods for those working for it are very much dependent on the financial success of such activity.
>
> Unfairness was particularly evident in the film's treatment of the Faroese case in defence of pilot whaling. This was presented through the medium of hidden cameras and microphones; techniques which had been deployed to obtain incriminating testimony from people involved in illegal trade in endangered species in other films in the Animal Detectives' series (please note EIA newsletter reference to 'corrupt officials and crime syndicates'.) We are surprised that such methods were deemed appropriate for a film dealing with a perfectly legal activity which takes place in an open society (*ibid*: 1-2).

In the program, EIA claimed that the Faroese were wealthy enough to forgo the free share of meat and blubber each individual received from the pilot whale hunts. Film clips showed the inside of the supermarket in Tórshavn with a voice-over by EIA 'undercover agents' informing viewers that the Faroese had all the trappings of a modern 'twentieth century economy' and that "lots of people are shopping, so they've obviously got money to buy" (Animal Detectives—Pilot Whaling In the Faroe Islands, 1995). Hoydal responded:

> What was noticeably absent was a look at the indicators by which a country's economic health is usually measured. Reflecting the unprecedented economic difficulties face[d] by the Faroe Islands in recent years, such statistics show a sharp fall in GDP, high levels of unemployment and foreign debt. The failure to cite such quantifiable evidence did not prevent the film baldly concluding that 'the meat is no longer needed by the people' (Hoydal 1995:6-7).
>
> Hoydal asserted that the program also:
>
> ...sought to paint a picture of a society ruled by intimidation. An assertion was made that it was 'very difficult for people to speak out because it is such a small community.' The presence in this small community of 8 political parties and an even greater number of religious denominations is a telling indication of the diversity of opinion and mutual tolerance that exists within it (*ibid*: 8).

The program served as a catalyst for the large letter writing campaign against the pilot whale hunt, as it enabled EIA to get the bloody images of the pilot whale hunt into the homes of millions living in postdomestic Britain in an attempt to shock many of them into action.

The Faroese people did not oppose a critique of their pilot whaling, they opposed the way in which it was being undertaken and the way that they were denied a voice in the discussion. In 1994, over 10,000 Faroese people signed an open letter to the Pilot Whale Campaign, which was delivered to EIA headquarters in London. Again, in 1995 a petition was signed by 8,000 readers of a Faroese newspaper (*Sosialurin*) in which they protested about "the way... organisations claiming to be concerned about the environment and animal welfare... use false and misleading information in their continued persecution of the Faroese nation and the very basis of our existence" (*ibid*: 4). Hoydal commenting on the Faroese petitions stated:

> What both of these expressions of opinion illustrate is that EIA and like-mined organisations have not become unpopular merely because they campaign against whaling. Rather, what has aroused antagonism from the Faroese people is the nature of the campaign which they saw as being characterised by hypocrisy, grotesque misrepresentations of pilot whaling, including cynical manipulation of visual images, and vilifications of their nation and way of life (*ibid*: 5-6).

Commenting, some twenty years after the anti-whaling protests began against the Faroese, Hoydal stated that what was disappointing from his perspective was the pattern of behaviour he observed among many of the anti-whaling organisations. He said: "Many of them criticised Faroese whaling, sometimes justly, but they never came back to see how we responded and what improvements were made" (Hoydal 2006, interview).

**Sea Shepherd Conservation Society**

The Sea Shepherd Conservation Society, founded in 1977 by a disaffected founding member of Greenpeace, is a marine mammal animal-rights organization currently based in Friday Harbor, Washington State, U.S. The split between its founder Paul Watson and Greenpeace was not amicable, with Watson claiming "that today the 'Warriors of the Rainbow' are little more than money grubbing, publicity-grubbing hypocrites, self-victimised by size and inertia" (Scarce 1990:102).

Sea Shepherd was one of the first animal rights groups to target Faroese whaling. Initially, Sea Shepherd focused on anti-sealing campaigns in Canada in the late 1970s, and then, as detailed on the Sea Shepherd website:

> [I]n 1979, the Sea Shepherd hunted down and rammed the notorious pirate whaler the *Sierra* in a Portugal harbour ending its infamous career as the scourge of the seas. The success of the seal campaign and the ramming of the *Sierra* was the start of Sea Shepherd's historical 160 voyages over the next 2 decades, enforcing international laws where no law enforcement existed—on the high seas (Sea Shepherd 2006).

Sea Shepherd had developed a reputation for direct action protests after ramming the *Sierra* in Portugal. Watson criticised other anti-whaling groups for not undertaking such action. He complained that Greenpeace had raised "three or four million dollars to stop the killing of pilot whales in the Faroe Islands, yet never goes there" (Scarce 1990: 102). In the mid-1980s Sea Shepherd began to target the Faroese pilot whale hunt and to use it as one of its major campaigns. During 1985, a Sea Shepherd vessel entered Faroese waters with the objective of interfering with the pilot whale hunts and gaining media attention for its anti-whaling fund-raising campaign. During its campaign, Sea Shepherd's major reason for opposing the pilot whale hunt was its claim that "the Faroese pilot whale kill was considered a serious environmental threat" (Sea Shepherd 2000).

After a lapse of fifteen years, in 2000 a Sea Shepherd vessel was again back in Faroese waters for a second campaign entitled "Grindstop 2000." During this campaign, largely undertaken via its website aimed at an American audience, Sea Shepherd once again claimed Faroese whaling was a threat to pilot whales (Sea Shepherd 2000).

In 2002, Sea Shepherd was engaged in another campaign against the Faroese people and their whaling. This time they targeted German companies who were granted exploration licences by the Faroese Government to explore for oil deposits in Faroese waters. The campaign was designed to scare joint venture partners from entering into agreements with the Faroese for fear of a boycott of their products in European markets.

**The Whale and Dolphin Conservation Society**

The Whale and Dolphin Conservation Society (WDCS), a British-based animal rights organisation, also ran several large campaigns focused on Faroese pilot whaling (WDCS undated, 1997). In fundraising letters the WDCS warned readers:

> Every year, in the cold Atlantic waters, less than two hundred miles from Scotland a barbaric slaughter takes place. It is bloodthirsty, pointless and immensely cruel. It does not make money or feed a starving population [...] All this killing is for little more than fun. Innocent, harmless, gentle creatures are knifed and hacked to death for sport. To add insult to injury, a few cuts of cheap meat are lopped off and the rest is left to rot on the beach (WDCS 1994).

On its website the organisation again outlined its opposition to Faroese pilot whaling:

> In the absence of any substantial economic or nutritional need, the Faroese have increasingly turned to their 'history' and 'culture' to provide justification for the continuation of the pilot whale hunt. This, given their rather successful integration into world market economies brings them to a rather contradictory and paradoxical position. On the one hand, they claim to be an isolated, self-sufficient and traditional society, while on the

other, their standard of living depends on them being just the opposite.

The Faroes cannot have the best of both worlds. Cultures, just like economies, are susceptible and, indeed, are required, to change. This cannot be more clearly illustrated than by the less-than-traditional features that have come to characterise the modern pilot whale drive. High-powered boats and modern communications equipment have come to replace the traditional methods of the hunt (WDCS 2001).

## Summary of Criticisms of Faroese Whaling

Despite the height of the campaigns occurring during the late 1980s and 1990s, they have not ceased. Each year, one or another group launches or re-launches a campaign against Faroese pilot whaling, and with each campaign come new threats to the Faroese economy and cultural traditions. Each campaign reinforces the idea that the Faroese pilot whale hunt lacks a management regime, is an unsustainable activity, has no place in a modern world, and the Faroese are 'backward' and 'isolated' from cultural, political and economic influences from the rest of the world.

Because of this isolation, the Faroese are portrayed as a discrete isolate with their own brutal, antiquated and unnecessary practices. Views like this are linked to the hunting of pilot whales, which, belonging to the past, should be left in the past.

Some environmental and animal rights' organisations asserted that ancient community-based common property resource institutions and the modern industrial world cannot co-exist, and that there can be no complex relations between the two: it is either one or the other. This proposal is evident in statements like, "if they want to kill whales in the traditional way, that's fine by us, if nothing else about their way of life, significantly anyway, has changed" (WDCS 1994).

In their critical objections to Faroese whaling, many of the groups use terminology and concepts without providing detail of what is actually meant. A recurring theme seen in the anti-whaling literature is one where the environmental and animal rights' organisations emphasize the problems inherent in issues of sustainability, without defining their understanding or use of the term 'sustainability.'

Reasons used by environmental and animal rights' organizations over the past two decades to bring an end to Faroese pilot whaling include the following:

- Faroese history and culture provide little more than justification for unsustainable whaling (WDCS).
- Whaling is a sport and the killing is little more than fun (WDCS, EIA).
- A subsistence and modern economy cannot co-exist in one society (WDCS).
- There is no economic need to whale (WDCS, EIA).
- The hunt is unsustainable (EIA, Sea Shepherd).

- The use of modern technology threatens sustainability (*International Field-sports Magazine*).

This book addresses these criticisms, by examining the Faroese institution of the *grindadráp* within a broad political, economic, social and ecological framework in order to assess the *grindadráp* in terms of its ability to sustainably maintain itself and the renewable resource upon which it is based, *viz.,* the North Atlantic long-finned pilot whale, over time.

To assist in making this determination, the following questions will be examined:

1. What is the nature of Faroese history, society, economy and environmental management of whales through the concepts of political economy and common property resources?

2. Is there a common-property resource institution for sustainably utilising whales in the Faroes? And, if so, how does it operate?

3. Does this common property resource institution reflect essential design principles, identified in anthropological and common property literature as being essential criteria for sustaining both the resource and the common property resource institution over time?

# *The Location, the Resource and Research Methods*

*Existing popular descriptions of the Faroes and their people lack a cultural-historical perspective. It is clear that it is foreign eyes that have observed, and foreign pens that have written. Outward activities are described, but without insight into the inner life of the population, without insight into the features peculiar to the culture.* Jakob Jakobsen 1904 (translation in Sanderson 1992:67).

A brief narrative of fieldwork in Faroese villages clearly demonstrates that Faroese society today is akin to what Bulliet (2005:3) describes as a domestic society, one where people consider commonplace and acceptable the killing of animals for food and experience few moral qualms in consuming animal products. Also described are the research methods used to enable the 'outside researcher' to develop the social relationships needed to gain insights into the inner life of the population and the common property resource institution in use for the utilisation of whales.

## The Origins of the Faroe Islanders

The Faroe Islands archipelago is situated in the northeast Atlantic Ocean between Iceland, Scotland and Norway (Fig. 3). Little is known of the early period of settlement of the Faroes, and what is known is derived from early written accounts and the archaeological record. The first settlers in the Faroes were Irish monks, who were thought to have arrived, along with their sheep, in the 8$^{th}$ century from Eire (Ireland) or from the north of Scotland (Debes 1989). It is not known if the Irish monks left the Faroes by choice, or were forced to leave following the 9$^{th}$ century arrival of Norse people from western Norway, or from the Shetland and Orkney islands.

Dicuil, an Irish monk and scholar, made reference to the Norse settlement of the Faroes in his geography manuscript, *De mensura orbis terrae* (the Book of the Measurement of the Earth) written in 825 AD. Here, Dicuil briefly described the time it took to sail in favourable conditions to the northern islands of Britain (the Orkneys and Hebrides), but he also described:

> Another set of small islands, nearly all separated by narrow stretches of water in these for nearly a hundred years hermits sailing from our country Scotia (Ireland), have lived. But just as they were always deserted from the beginning of the world, so now because of the Northern pirates *('causa latromum Normannorum')* they are now emptied of anchorites, and filled with countless sheep and very many diverse kinds of sea-birds. I have never found these islands mentioned in the authorities (Tierney 1967, cited in Debes 1989:24).

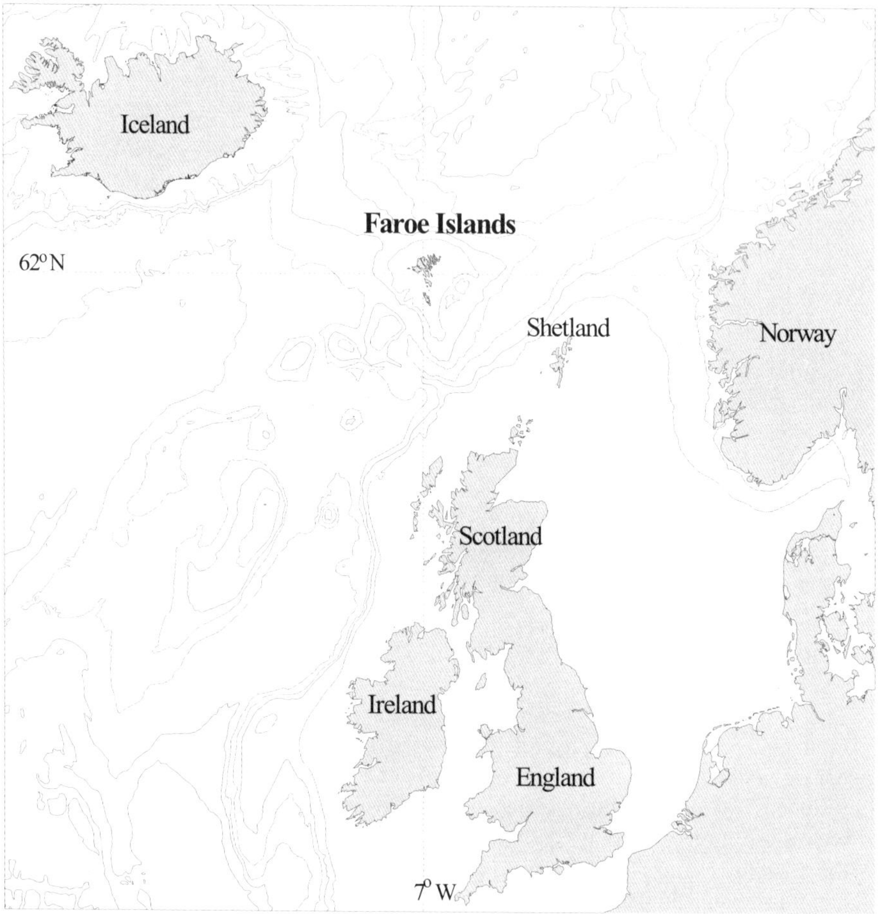


**Figure 3.** *The location of the Faroe Islands in the northeast Atlantic Ocean.* (Map by Lise Ofstad)

The 'small islands' described by Dicuil, cannot be Shetland, for they had already been settled for thousands of years, nor Iceland, which he describes elsewhere in his manuscript (Debes 1989). The small islands must surely be the Faroes, for when looking at a map of the Faroes one is struck by the archipelago's most striking geographical feature, the northwest–southeast direction of the narrow channels that separate the islands from one another. Furthermore, the Faroes remains a place with large numbers of sea-birds.

Archaeological evidence also establishes the date of Norse settlement of the Faroes at approximately the 9th century (Arge and Hartmann 1989). Scattered archaeological and linguistic evidence suggests the first Norse settlers came to the Faroes mostly from western Norway, but to some extent also by way of Shetland and the Orkneys which were already under Norse control (Wylie 1987).

Hansen (2003) concluded that the Scandinavian emigrants to the Faroes brought with them their identity in the form of the very standardised architecture of the classic Viking longhouse. Additional material culture, demonstrated through the characteristic ringed pins, armlets and finger rings of jet and the adoption of the quern and mill technology, reflected the insular identity developed with contacts from the Celts in Britain and Ireland.

Archaeological discoveries indicate the early settlers were primarily farmers, raising sheep, cattle, and pigs. They used horses and grew crops to a greater extent than is possible today. The finds in middens indicate that from early settlement times, some fishing and hunting of pilot whales and seals provided supplementary sources of food. The current population, as discussed below, is descended from these Viking settlers as shown in language and other customs.

**The Concept of Indigeneity and Aboriginality**

Does being the first permanent settlers on the islands make the Faroese indigenous or Aboriginal? This question is asked because the concept of indigeneity is a point of demarcation between permissible and impermissible whaling within the international debate on whaling. For example, as discussed above, the IWC categorises whaling into two groups: commercial whaling, which it does not permit, and aboriginal subsistence whaling, which it does. The IWC has defined Aboriginal subsistence whaling in the following way:

> Aboriginal subsistence whaling means whaling, for purposes of local aboriginal consumption carried out by or on behalf of aboriginal, indigenous or native peoples who share strong community, familial, social and cultural ties related to a continuing traditional dependence on whaling and on the use of whales.
>
> Local aboriginal consumption means the traditional uses of whale products by local aboriginal, indigenous or native communities in meeting their nutritional, subsistence and cultural requirements. The term includes trade in items which are by-products of subsistence catches.
>
> Subsistence catches are catches of whales by aboriginal subsistence whaling operations (IWC 1982:5).

The distinction between these two categories of whaling, *viz.* commercial and aboriginal subsistence, is blurred, with both categories sharing some commonalities. As Freeman indicates:

> Both [aboriginal and non-aboriginal] are involved, *inter alia*, in subsistence (i.e. they sustain community traditions and identity by engaging in a range of whaling-associated activities that are highly valued in their own communities); both operate family- or community-based operations; both exist within an economy that involves subsistence and commercial activities; and both

> contribute significantly, through their whaling, to the economic viability and social vitality of their communities (Freeman 2001:138).

The former Secretary to the IWC clearly acknowledged the problem with the definitions within the IWC when he asked: "What is the difference between feeding your family directly with whale meat and selling the whale meat in order to obtain food for your family?" (Gambell 1993, cited in High North Alliance 1994:3). Indigenous hunters who fall within the IWC definition of aboriginal subsistence whaling, and who benefit from being able to whale under IWC processes, assert that the definition is no more than a form of inverted racism. Leif Fontaine, the Chairman of Greenland's fishers' and hunters' organization, KNAPK (*Kalaallit Nunaanni Aalsartut Piniartullu Kattufiat*) stated that:

> To deny us the right to hunt commercially and trade in our products is to lock us in a museum case. We are a people living in the 21st century, not an exhibit from the British Museum... Commerce is part of the concept of sustainable development, and it is our right to participate in this. To limit aboriginal whalers to a non-commercial regime is nothing other than neo-colonialism (Fontaine, cited in High North Alliance 2000:9).

Other commentators assert the definition is no more than a 'concept of imperialism' (Kalland 1994:39), and illustrates the problem with the term as it is used within the IWC:

> Being in charge of their own destiny, what separates the Greenlanders as a people from the Faroese, apart from the colour of their hair? Both have obtained home rule within the Danish Realm, so politically there is hardly any difference. Both people are affluent in the global context. Both catch whales, but whereas the Faroese catch pilot whales which are consumed locally, much of the minke and fin whale meat taken by Greenlanders reach the market. Yet, Greenland's whaling is termed subsistence, while the Faroese is condemned for not being 'necessary' (Kalland 1994:40).

As the IWC itself notes, "any whaling activity has its own culture, and is deeply rooted in the history of the nation; as both types of whaling involve the same interaction between man and whales as a resource then the same principles and management objectives should apply" (IWC 1982:3).

The definition of aboriginal subsistence whaling thus being ambiguous, can be interpreted in different ways for different purposes. For example, in 1995 the Makah people of the northwestern U.S. requested a quota of five gray whales (to be taken from a population estimated to number 26,300), after a 70-year hiatus of their whaling operations (IWC 2007). New Zealand opposed the request within the IWC and claimed that the Makah had no 'continuing traditional dependence' on whales and therefore the Makah should not be granted a quota under the aboriginal subsistence whaling category of the IWC (McLay 1995, 1996). Here, New Zealand asserted its own definition of aboriginal subsistence whaling which it also promoted within the IWC debate. New Zealand considered:

> ... only the legitimate needs of *authentic* indigenous claimants should be recognised within applicable international law and IWC frameworks. On that basis, aboriginal subsistence whaling can be accepted, provided that requests meet IWC criteria, which are that:
>
> - whale meat must be obtained to meet an established cultural tradition and nutritional need;
> - whale meat must be consumed locally;
> - whale meat must not be utilised for commercial purposes; and
> - aboriginal subsistence whaling must not threaten the ecological viability of the stock (DOC 2006: 22, emphasis added).

It is not only within the IWC that the definition of indigenous (aboriginal) is ambiguous. The International Work Group for Indigenous Affairs (IWGIA), an independent international membership organisation states, for example:

> If the word 'indigenous' refers to the 'origins,' all human beings are indigenous in some way. However, in the sociological and political vocabulary (and evermore often in the juridical usage as well), the term 'indigenous' is employed as a reference point to groups of the population that occupy a determined position in society as a result of specific historic developments (Assies and Hockema 1994:14).

The IWGIA also states that:

> The current use of the concept of 'indigenous' implies the idea of original occupation of a given territory. In this sense, without doubt indigenousness is an ambiguous category since in most cases original occupation cannot be actually documented. No one knows for sure who were the original occupants of a given territory. Those who nowadays claim to be 'indigenous' may in the distant past have replaced anterior occupants (*ibid*: 16).

The concept of 'indigeneity' suggests a historical continuity between the original population and those who presently identify themselves as direct descendants from that population. This continuity can be genetic (through biological reproduction) and cultural (maintenance of cultural forms such as language or religion that derive directly from the original group) (*ibid*: 17).

The special rapporteur for the United Nations proposed yet another definition:

> Indigenous communities, peoples and nations are those which, having a historical continuity with pre-invasion and pre-colonial societies that developed on their territories, consider themselves distinct from other societies now prevailing in those territories or parts of them. They form at present non-dominant sectors of society and are determined to preserve, develop and transmit to future generations their ancestral territories, and their ethnic identity, as the basis of their continued existence as peoples in accordance with their own cultural patterns, social institutions and legal systems (Martínez Cobo 1983:30).

Thus the concept of indigeneity can be a highly ambiguous term. Some who are categorised as indigenous and would appear to benefit from the IWC definition clearly reject it. How the concept 'aboriginal whaling' has been employed within the IWC serves no purpose, particularly in regard to any analysis of a discussion on Faroese whaling (*see* Akimichi *et. al.* 1992 and Doubleday 1994). For these reasons, the concept of indigeneity is set aside in further discussion of Faroese whaling, as it offers no conceptual assistance in the discussion that follows.

## The Land

It is necessary to understand the severity of climate and fragility of the land in order to understand interaction of the Faroese with their environment. The Faroes are situated in a highly productive ocean environment. The archipelago is 113 km long and 75 km wide with a landmass of 1,399 km$^2$ composed of 18 islands (Fig. 4). These islands vary in size from one to 374 km$^2$, with no location more than 5 km from the sea. The average height of land above sea level is 300 m, with the highest peak, Slættaratindur, rising 882 m above sea level.

The Faroe Islands archipelago was formed through volcanic activity, when successive lava flows extruded from large fissure eruptions in high ridges rising from the ocean floor (Rasmussen 1982). Lava was built up in successive layers interspersed with volcanic ash, giving the islands their pancake-like appearance. The land was then slowly eroded by the forces of wind, water, frost, and glaciers with the coastline shaped by the actions of the ocean.

The land is deeply indented by narrow arms of the ocean and separated by passages whose strong tidal currents can make navigation hazardous. The coastline varies from small sandy beaches to enormous sea cliffs rising up into the mists. Within the fjords and sounds, the coast is protected from the waves, with the sides of the fjords sloping toward the water.

The climate is characterised as cool, wet and windy with many overcast days. Frost occurs as often as 68 days a year during the months of December through to March. The air temperature rarely remains below 0°C for a full twenty-four hours, with the average temperature ranging from 3°C in the winter to 11°C in summer. Precipitation occurs over 281 days of the year and averages 1,461 mm annually, although there are localised variations. Although snow falls, on the average, 44 days a year, it rarely remains for more than two weeks. The relative humidity varies little and the yearly average is 83 percent. Skies are generally cloudy, and fogs are frequent and often localised, although seldom remaining for more than eight hours (Rutherford 1982).

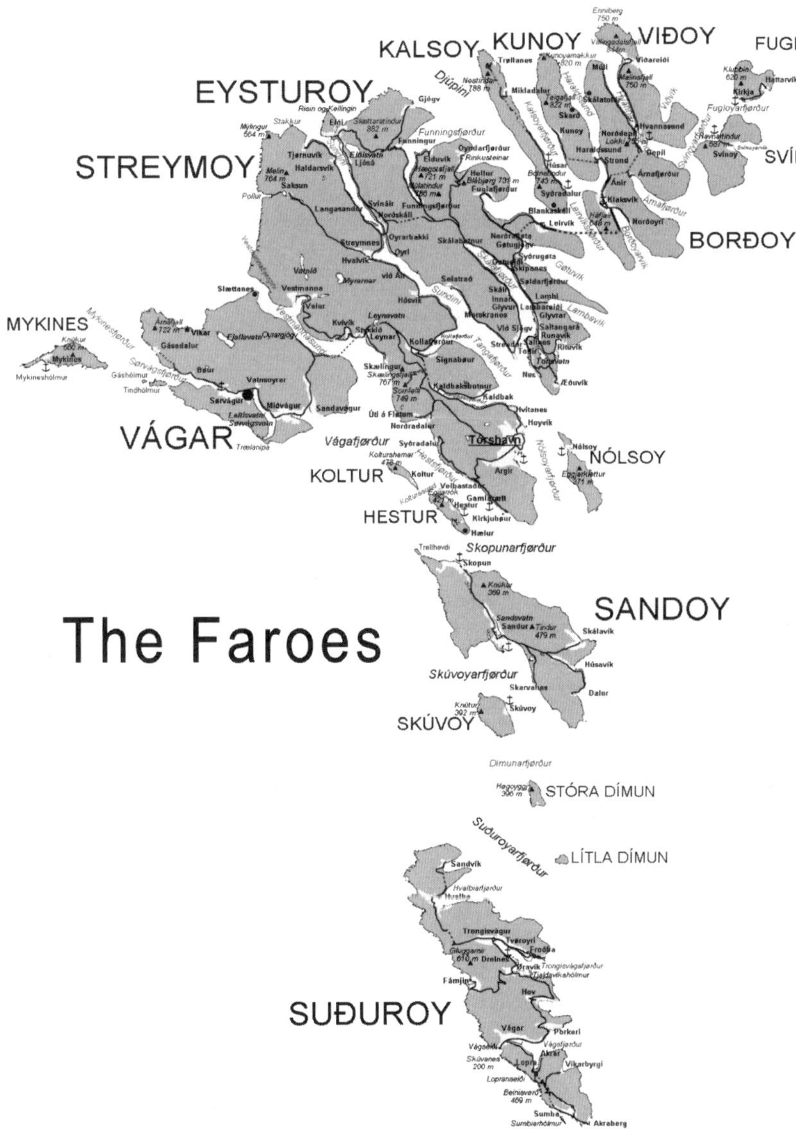


**Figure 4.** *Map of The Faroe Islands* (mappery.com/map-name/Faroe-Islands-Map).

The islands have no major river; the landscape is rocky and rugged with a thin fragile covering of soil that supports grasses and mosses. No forests grow and trees are rare, occurring mostly around human settlement. There is very little flat land, and only six percent of the total land area is arable, which together with a short growing season, allows very little agriculture to be practised on the islands. While barley was once grown on the Faroes, it never ripened during the short summer and had to be reaped green and dried over peat fires before it could be used. Hay is the main crop grown today; it is cut, usually by hand, during the summer, partially dried in the fields and then made into silage to feed the sheep over the winter.

Sheep are by far the most numerous domestic animal on the Faroes and are thought to have been on the islands since the settlement by Irish monks over one thousand years ago. The name 'Faroes' (*Forøya*) means 'sheep islands' and the oldest breed of sheep were evidently small and dark, resembling St. Kilda sheep. The last of this old breed were killed off on Lítla Dímun in 1868 (Wylie 1987). Many Faroese households, especially in the smaller villages, keep a few sheep for their own subsistence needs. The Faroes, with its poor soils and mountainous landscape, has a population of only 70,000 sheep, or about 1.5 animals per person, making the importation of sheep meat (lamb), largely from New Zealand, necessary to meet the high demand for sheep meat.

## The Sea

The North Atlantic Current greatly influences the local climate and provides the resources on which the Faroese economy is built. The sea is the provider of life to the Faroese, and features in stories, songs and art. Bogi Hansen powerfully describes the importance of the sea to the lives of people living in the northeast Atlantic:

> The sea separates and the sea unites. The sea feeds and the sea kills. For the West Nordic countries the sea has, to a large extent, determined how man could exist and cultures develop. With its immense distances and violent conditions, the sea has separated the West Nordic societies from one another and from the rest of the world. At the same time, it has provided a means for transport that through the ages allowed contact, for good or bad. And, the sea has made survival possible in areas where the land was too poor. The West Nordic societies, especially the hunters, have lived from the sea. It has provided them with food and other necessities; but not without cost. The West Nordic seas are rich and sustain an abundant animal life, on which man has survived, but they are also harsh. Waves, currents, storms, icebergs may quickly turn a hunting voyage into a wet funeral (Hansen 2002:112-114).

The Faroes sit atop the Faroe Plateau, which extends to depths of approximately 200 m., and along its southwest and north sides sloping steeply into the deeper waters of the Faroe Bank Channel (500–1,000 m) and the Norwegian Basin (500–3,000 m). The Iceland–Faroe Ridge extends north-west from the shelf area of the Faroe Plateau toward Iceland. A distinct topographic feature on the seabed to the southwest of the islands is the Faroe Bank, which rises to depths of less than 100 m above the Faroe Bank Channel. The edge of the shelf to the east of the islands is less well defined, with its progression into deep waters following a more gentle slope. The Faroe–Shetland Channel extends below 500 m to depths of over 1,500 m to the southeast of this area (Atlantic Margin Group 1997).

The Faroes lie near the transition of temperate Atlantic to Arctic environments at the confluence of two very different water masses: the East Icelandic Polar Current and the other of tropical origin, the North Atlantic Current. It is these water masses which are of primary importance for both the abundance of marine life and the climate of the islands. The sea nurtures an enormous number of different organisms, from tiny microscopic creatures to the huge blue whales. In the sea, as on land, the plants are the basis for all living organisms. One publication describes the seasonal arrival of light and the burst of plankton, which supports the complex food chain on which the Faroese depend:

> During the spring, day length increases and the thermal gradients together with reduced wind stress begin to stabilise the upper water column. These factors, together with a supply of nutrient rich water from turbulent mixing during the winter storms, provide good conditions for phytoplankton growth. The rate of growth in the phytoplankton is particularly rapid at the start of the spring 'bloom,' but is quickly controlled by a reduction in nutrient levels, as they become incorporated in the plankton population and by increases in zooplankton grazing pressure. The spring bloom of phytoplankton occurs around early May in Faroese waters, but its timing and intensity will vary with the conditions pertaining to a particular year (Atlantic Margin Group 1997:12).

Due to this high productivity of plankton in the waters surrounding the Faroes, the adjacent sea supports a large number of marine mammals and a vast number of fish species. Important commercial species include: Atlantic cod (*Gadus morhua*), saithe/coalfish (*Pollachus virens*), haddock (*Melanogrammus aeglefinus*), halibut (*Hippoglossus hippoglossus*), redfish (*Sebastes viviparous*), herring (*Clupea harengus*), blue whiting (*Micromesistius poutassou*), capelin (*Mallotus villosus*), sand eel/sandeel (*Hyperoplus lanceolatus*), monkfish (*Lophius piscatorius*), and various species of shrimp.

Not only does the ocean support large and diverse fish populations, but also huge migratory bird populations. Seabirds include: fulmar *(Fulmarus glacialis*), gannet *(Sula bassana*), kittiwake *(Rissa tridactyla*), Arctic tern (*Sterna paradisaea),* guillemot *(Uria aalge*), puffin (*Fratercula arctica*), Leach's petrel (*Oceanodrama leucorhoa*), storm petrel (*Hydrobates pelagicus*),

Manx shearwater (*Puffinus puffinus*), sooty shearwater *(Puffinus griseus*), and great skua (*Stercorarius skua*) (Skov *et al.* 1995). Many of these species have long histories of exploitation by the Faroese in their subsistence economy and have been essential to the survival of the population during harsh periods of climate and bleak economic periods such as that created by the Danish trade monopoly from the 16$^{th}$ to the mid-19$^{th}$ centuries and the economic collapse in the late-20$^{th}$ century. Today, approximately 100,000 seabirds are taken each year for local consumption. The harsh environment, small land mass, and fragile soils means that permanent settlement of the islands has been a challenge to which the Faroese have responded by utilising diverse and locally-abundant marine resources.

## Natural Resources

As discussed above, a distinctive feature of managing whales is that many people think of them as 'special creatures,' also referred to as 'charismatic megafauna,' and as such, requiring a unique management regime often quite unlike that applied to most other wildlife species. Consequently, being 'special,' whales are not considered a potential human-food resource by people holding that particular belief.

However, this study considers whales, together with all other forms of wildlife, to be natural resources, defined as "those components of an ecosystem which provide goods and services useful to man" (Gibbs and Bromley 1989:22). Natural resources can exist in two forms: as stocks, such as coal or mineral deposits, or flows such as water, forests, fisheries or wildlife:

> For stock resources the physical quantity available for use is more or less fixed: what is used now will not be available later. Stock resources are therefore exhaustible and non-renewable and it is possible to identify an optimum rate of depletion based on stock resources. Flow resources are renewable and can be managed to yield goods and services sustainably (Gibbs and Bromley 1989:23-24).

According to this definition, all whales are flow resources, and therefore have a critical zone below which a decrease in flow may not, or cannot, be reversed when the population is reduced below that critical minimum size. According to Gibbs and Bromley (*op. cit*), once the critical zone is approached or has been reached, the flow of the resources dramatically diminishes or ceases completely, and the resource itself becomes threatened. Throughout history, users of flow resources, in order to maintain their own survival, have needed to ensure that these critical zones are never reached. This is a difficult and complex task that requires resource users to act collectively in their use and management of such resources to avoid resource over-exploitation and eventual possible disappearance. Knowledge about the resource must be developed by its users, including an understanding of resource population behaviour, its feeding requirements and its predators, breeding behaviour, its migratory behaviour, and how best to capture it.

**The Long-finned Pilot Whale**

The long-finned pilot whale (*Globicephala melas*) has several other names, including the caa'ing (or 'calling') whale in the Orkneys, and the pothead whale, blackfish or bagfin in England. In Faroese it is known as the *grindahvalur*. The head is bulbous and melon shaped, which is reflected in the name *Globicephala*.

The long-finned pilot whale is a toothed whale belonging to the family *Delphinidae,* so in a scientific sense, it is a dolphin, not a whale. Pilot whales are dark brown to black in colour, sometimes with grey bands near the eyes; they also have a distinguishing greyish-white, anchor-shaped mark on their throats. The dorsal fin is set forward (in the front half of the back), low but wide at the base. In females, it is more triangular, whereas males have a more sickle-shaped dorsal fin. The pectoral fins, from which the long-finned pilot whale gets its name, are very long, on average 27 percent of the total body weight. Males are larger than females, reaching a length of 6.3 m and a weight of 2.5 tonnes, compared to 5.5 m and 1.5 tonnes for females. Pilot whales' breathing pattern consists of a few short breaths before diving for a period ranging from five to 10 minutes. They are known to dive in search of food to a depth of at least 600 m. Pilot whales socialise in schools of up to several hundred individuals (Watson 1981; Amos *et al.* 1993; Bloch *et al.* 1993; Fontaine 1998).

Male pilot whales reach sexual maturity at age 10 to 15 years (Desportes *et al.* 1994), while females mature at age 5 to 15 years, with both sexes reaching physical maturity at age 25 to 30 (Bloch *et al.* 1993; Martin and Rothery 1993). Gestation lasts approximately 12 months, with female fertility declining with age so that whales older than 40 years rarely become pregnant (Martin and Rothery 1993). Recent studies indicate that the highest fertility rates occurred in 10-year-old animals; individuals more than 32 years age were considered to be reproductively senile, even though there was evidence that they continue ovulating and lactating (NAMMCO 1998b).

Like many species of toothed whales, pilot whales are highly social animals and usually occur in schools of 10 to 200 whales, with larger schools occurring more rarely. Schools generally consist of related females with offspring of both sexes and all ages, with adult females outnumbering adult males. However, male-only schools and schools consisting mainly of males have been recorded (Desportes *et al.* 1992). Genetic evidence indicates that while males remain in their natal school, they do not father calves in that school. Therefore, it is believed that mating must occur when two or more schools meet (Amos *et al.* 1993). Many pilot whales carry evidence of fighting scars from teeth marks, which can be clearly seen when whales are beached. Long-finned pilot whales have few predators besides humans, but may be preyed upon by killer whales (*Orcinus orca*) and large sharks.

Pilot whales are opportunistic feeders that exploit any locally-abundant prey; however, their primary food species is squid. There is a slight difference in the diet of pilot whales between those in the northeast and northwest Atlantic, with northeast Atlantic pilot whales feeding mainly on squid (*Gonatus* spp. and *Todarodes sagittatus*) but also taking fish, including greater argentine (*Argentina silus*), blue whiting (*Micromisistius poutassou*), Greenland halibut (*Reinhardtius hippoglossoides*), and pandalid shrimp (Desportes and Mouritsen

1993). In the northwest Atlantic, squid also dominate the diet, target species being short-finned and long-finned squid (*Illex illecebrosus* and *Loligo pealei*), but pilot whales in that region also eat Atlantic mackerel (*Scomber scombrus*) and herring (*Clupea harengus*) (Mercer 1975; Gannon *et al.* 1997a,b).

Long-finned pilot whales may be one of the most important marine-mammal predators in the offshore waters of northwestern North America (Gannon *et al.* 1997a,b). In the central North Atlantic around Iceland, pilot whales are considered to be the fourth most important marine mammal predator in the marine ecosystem, consuming 200 to 300 thousand tonnes of fish, and 900 to 1,100 thousand tonnes of squid annually (Sigurjónsson and Víkingsson 1997). Much of this consumption is concentrated on species presently of little direct importance to commercial fisheries, so competition with fisheries is currently considered minimal.

The distribution and stock definition of two species of pilot whales in the North Atlantic—the long-finned and short-finned pilot whales—were long confused, but are now well-defined (Sergeant 1962). Their distribution is described as 'parapatric' (meaning the ranges of the two species abut, but do not overlap). Both species have a circumglobal range in the northern and southern hemispheres, with the short-finned pilot whale restricted to tropical and sub-tropical waters, and the long-finned pilot whale found in polar to temperate waters (Rice 1998).

Long-finned pilot whales are an oceanic species, occurring far offshore as well as in the coastal areas. In the North Atlantic they are widely distributed, from 35° to 65°N in the northwest Atlantic, and from 40° to 75°N in the northeast Atlantic (ICES 1996). There appears to be a gap in distribution in the area south of Greenland (the Irminger Sea), however this area has not been surveyed extensively. There is no indication that pilot whales undertake extensive seasonal migrations; however their distribution does change on a seasonal basis, probably in relation to the abundance of squid, which is their principal prey (Payne and Heinemann 1993; Zachariassen 1993).

Evidence suggests there are at least two stocks of long-finned pilot whales in the North Atlantic. First, there are morphometric (body shape) differences between pilot whales caught in the northwest and northeast Atlantic (Bloch and Lastein 1993), indicating that pilot whales from the two areas are unlikely to be from the same stock. Second, the marked decrease in pilot whale numbers off Newfoundland from 1947 to 1972 had no apparent effect on pilot whale abundance elsewhere, indicating that there is probably little or no exchange between that particular area and other areas (Mercer 1975; Nelson and Lien 1996). In addition, there may also be differences over smaller spatial and temporal scales suggested by significant differences in pollutant concentrations (Aguilar *et al.* 1993; Caurant *et al.* 1993) and parasite burdens between schools of whales landed in the Faroes at different times and locations (Balbuena *et al.* 1995). This suggests that these schools spend different proportions of their time in different areas, which again may be indicative of stock differences.

In 1997, the Scientific Committee of the North Atlantic Marine Mammal Commission (NAMMCO) concluded that, based on the evidence noted above, it was likely that there was more than one stock of pilot whales in the North

Atlantic, and more than one stock subject to harvesting in the Faroe Islands (NAMMCO 1998a).

The NAMMCO Scientific Committee has estimated the population of long-finned pilot whales derived from a number of sighting surveys carried out in 1987, 1989 and 1995 to be 778,000 animals (NAMMCO 1998b), a figure almost identical to the 780,000 estimate made by the IWC Scientific Committee (IWC 2007).

In regard to the sustainability of the Faroese catch of pilot whales, the NAMMCO Scientific Committee concluded that the effects of historic and present catches in the Faroes have had negligible effect on the long-term trends in the pilot whale stock. The Scientific Committee noted that an annual catch of 2,000 individuals in the eastern Atlantic corresponds to an exploitation rate of 0.26 percent of the present best estimate in abundance of 778,000 pilot whales in the northeast Atlantic (NAMMCO 1998a).

While noting the recommendations for further research, the NAMMCO Scientific Committee considered that it had now completed its work in addressing the Council's requests for advice on the long-finned pilot whale and the sustainability of the Faroese catch of pilot whales. The conclusion was based on the Scientific Committee's assessment of all available information, all of which had been thoroughly reviewed by the International Council for the Exploration of the Sea (ICES[1]) Study Group on Long-finned Pilot Whales. A catch of little more than one-quarter of one percent of the total stock indicates that the flow (catch) of the long-finned pilot whale resource is sustainable. Although the notion of 'sustainability' is highly contested, nonetheless it is generally accepted that sustainability is an objective that all users of flow resources should aim to achieve. The World Commission on Environment and Development (WCED) defined sustainability as "development that meets the needs of the present without compromising the ability of future generations to meet their own needs" (WCED 1987:8). Sustainability, as used here, "is a process and includes ecological, social and economic dimensions... sustainability implies not challenging ecological thresholds on temporal and spatial scales that will negatively affect ecological systems and social systems" (Berkes and Fölke 1998:4).

Whether or not resources are capable of sustaining communities of people with the resource itself being sustained indefinitely over time depend on numerous factors which are further discussed below.

---

[1] ICES was formed under the 'Convention for the International Council for the Exploration of the Sea' in 1964. ICES is the organization that coordinates and promotes marine research in the North Atlantic. Its membership comprises Belgium, Canada, Denmark, Estonia, Finland, France, Germany, Iceland, Ireland, Latvia, Lithuania, the Netherlands, Norway, Poland, Portugal, Russia, Spain, Sweden, the United Kingdom, and the United States of America. Each of the member countries elects two delegates who represent their country on the ICES Council. The ICES Council is the principal policy and decision-making body of ICES. A number of countries have affiliate status with ICES including: Australia, Chile, Greece, New Zealand, Peru, and South Africa (ICES 2005, 2006).

## Research Methods

This study was based on social research because the research questions are about social interactions in Faroese society. While social research transcends common sense and everyday views of the world, it nevertheless contains political and ethical issues that cannot be avoided. Such research is never neutral, for there is always a social context within which the research is situated. However, this does not mean that social research cannot be objective (Williams 2003).

The goal of this study of Faroese pilot whaling is to describe and understand what was taking place in regard to the pilot whale hunt at the beginning of the 21st century. To reach that objective, qualitative methods were considered appropriate in order to examine the complexity of relationships linking history, culture, and environment to local and regional management institutions. As Agnew and Pyke state: "A holistic approach will generate a rich, diverse and complex informational foundation from which a host of valuable speculations may emanate, some of which may be amenable to subsequent investigation using quantitative methods" (Agnew and Pyke 1994:168).

A major component of the research reported in this book was undertaken through fieldwork carried out on the Faroe Islands in 2001-2002 and 2006; methods included direct observation and personal engagement with community members in their daily activities. The goal of all such activities is to attempt to gain, to the greatest extent possible, an insider's perspective on Faroese life. The researcher embraces an activist or social constructionist perspective on social life, not seeing people as a neutral medium through which social forces operate, nor social meanings merely as something 'out there' to be observed, but rather, holding that people create and define the social world through their interactions. Thus human experiences are filtered through a subjective sense of reality (Lawrence-Neuman 1994), and fieldwork is not a passive activity in which a researcher arrives and merely looks, listens, experiences and records these observation. Fieldwork then, is an active process where the researcher becomes engaged with, and finds a place within, the community, often through 'participant observation,' which was once the preserve of anthropologists but has now become more widely adopted in other social science disciplines, including sociology, policy science, and human geography.

A brief narrative description of the author's fieldwork within the wider North Atlantic region and the Faroes is included to provide research transparency in order to address any issues of personal bias that may be perceived because of his participation in a number of whale and dolphin hunts, consumption of the products of those hunts, and previous involvement with whaling communities elsewhere. Prior to undertaking this fieldwork, the author travelled to Reine in the Lofoten Islands, 250 km inside the Arctic Circle and three hours by ferry from the coastal city of Bodø in northern Norway. Lofoten is the home to many of Norway's fishers/minke whalers, and Reine is where the High North Alliance (HNA) is based; the HNA is an advocacy organisation whose objective is "to secure the rights of whalers, sealers and fishermen to harvest renewable resources in accordance with the principles of sustainable management" (Hallenstvedt and Blichfeldt 1995:39). Communities allied to and served by the HNA are located in Norway, Greenland, Iceland, the Faroes, and Canada. The author spent four months liaising with representative individuals and groups within whaling communities located around the North Atlantic, and assisting in the production of informational material for the 53rd IWC Annual Meeting held in London in July 2001. While in Norway, the author interviewed a number of Norwegian minke whalers and examined whaling boats, hunting equipment and facilities used to process the meat and blubber once it was landed. This provided the opportunity to develop an understanding of a small-type commercial whaling operation and the regional setting for North Atlantic whaling.

Attending the 53rd IWC meeting with members of the HNA also provided the opportunity to observe the dynamics of the international institution (the IWC) charged with managing the use of the world's large cetaceans, and the opportunity to view the IWC both in the public arena within the plenary sessions, and behind the scenes in meetings of the pro-whaling nations and allied sustainable use organizations

In addition to mentioning these particular activities that provided a broader understanding of other forms of whaling, the author's employment as a policy analyst at the Treaty of Waitangi Fisheries Commission (*Te Ohu Kai Moana)* in New Zealand is also relevant. Here, the author worked with Maori[2] who were asserting their treaty rights to access beached whales, a situation that led the author to be labelled 'an apologist' for whaling by one of the New Zealand Government advisers and commentators on whaling politics.

This research on Faroese pilot whaling was undertaken following an invitation to engage in such research from the *Grindamannafelagið* (Faroese Pilot Whalers' Association). Despite receiving this invitation, the author had no formal or informal contract with the *Grindamannafelagið* and was free to independently design research objectives and questions. The *Grindamannafelagið* advised that they wished to contribute to available knowledge that could inform on-going discussion in the international community about their management institutions, culture, history and use of pilot whales. With this in mind, it was the view of several executive members within the *Grindamannafelagið* that the literature on Faroese pilot whaling in the English language was limited, and this was considered reason enough to encourage research by foreign researchers.

---

[2] Maori were asserting their Aboriginal and *Treaty of Waitangi* (1840) rights to traditional wildlife resources over which they had exercised long-term use and management rights. Besides fisheries, settled via the *Treaty of Waitangi (Fisheries Claims) Settlement Act*, Maori sought management rights over other traditional wildlife resources such as *Toroa* (albatross; *Diomedeidae), Kereru/Kukupa* (New Zealand wood pigeon, *Hemiphaga novaeseelandiae), Tohora/Paraoa* (whales), including minke whale (*Balaenoptera bonaerensis*), right whale (*Eubalaena australis*), and sperm whale (*Physeter macrocephalus*).

**Overview of Fieldwork in Faroese Communities**

The purpose of providing a brief overview of Faroese village life is to describe how I, as a researcher, participated in Faroese life and to demonstrate how the activity of whaling is set within a wider subsistence society in which many people living in the Faroes participate. What can be clearly seen from this narrative is that whaling is a real life event involving real people and their very real use of natural resources to provide food. In the Faroes, whaling is not an extraordinary activity but something quite commonplace.

Fieldwork for this book was undertaken in the Faroes on two separate occasions: the first, from June 2001 to May 2002, and the second, in July and August 2006. During the summer, much time was spent working with an extended family in Leirvík, in the arduous task of making hay for the winter food supply for a small number of sheep that the family kept. Haymaking was undertaken by hand, except for a small, motorised mower, not much larger than an ordinary lawn mower, and in some instances, a small motorised rake. Despite the smallness of the narrow peat fields, cut with numerous drainage ditches, the haymaking was a labour-intensive task. With short periods of constant sunlight and regular summer showers, the hay had to be constantly raked and turned in order to dry. At night it was raked into small piles and tied down to stop it being blown away by the frequent wind, and next morning it was spread out once again to continue drying. After a week or so when it showed signs of partial dryness, it was raked and pushed into woolsacks before being hauled up the sloping fields with the help of a pulley system rigged to a car up on the road, where it was compressed in a small concrete silo by placing a large plastic bag filled with water on top of the hay, after which it would eventually be used as sheep fodder during the winter.

Haymaking was a very social time, with many family members assisting when time allowed. As the weather dictated, haymaking was occurring on the majority of islands within a few short weeks. The cutting of the grass to make hay accentuated the patchwork pattern of the fields, making it startlingly obvious that the old Faroese land tenure system of infields (*bøur*) and outfields (*hagi*) were still in use today just as much as they have been since the settlement of the islands in the 9$^{th}$ century (Fig. 5).

Faroese villages are always clustered in the infield, which is divided into small plots used for crops, hay, and winter grazing; a stonewall usually surrounds the infield. The outfield is uncultivated, except in some isolated enclosures called *traðir*, and are used for the summer grazing of sheep. "The basic principle of land use, still in existence today, is that ownership or leasehold of each *mørk*[3] of outfield entails proportional rights over certain plots of infield, and to similar rights over sections of the shore and the birdcliffs" (Wylie 1987:31).

---

[3] The basic unit of value in land was (and is) the 'mark' (*mørk*, pl. *merkur*). There are about 2,400 *merkur* of land in the Faroes by traditional reckoning. A *mørk* is divided into sixteen *gyllin*; a *gyllin* is in turn divided into twenty *skinn*. The *gyllin* appears to be a Hanseatic unit that came to the Faroes in the late-15$^{th}$ century (Wylie 1987:31).

**Figure 5.** *The village of Viðareiði on the northern island of Viðoy, showing the hay-cutting system and the infields and outfields* (Kerins 2006).

Haymaking provided extended periods of time for me to participate in daily life and ask questions about the history of land tenure, the use of sheep, seabirds, whales, dolphins and other resources. The whole time hay was being made, a constant eye was kept on the sea hoping to sight a school of pilot whales.

The short breaks available during haymaking provided the opportunity to experience other traditional summer subsistence activities, such as puffin hunting on nearby Fugloy, the furthest northerly island in the Faroes. Fugloy is a small steep island inhabited by a small population of about twenty people, which marginally increases over the short summer period when other family members return from Tórshavn or other smaller villages around the islands or from Denmark. Fugloy has two small villages, Hattarvík and Kirkja, and is serviced by a small ferry making the daily one-hour trip from Hvannasund on Borðoy via Svínoy bringing supplies of fuel, milk, fruit and vegetables and sometimes tourists who occasionally make a sight-seeing visit—although rarely leaving the ferry.

Each alternate year the residents of one of the two villages on Fugloy have rights to harvest the puffins while they nest on the island during the summer. The steep cliffs where the puffins nest are reached via the ocean, unlike other fowling cliffs in the Faroes where men are lowered down on long ropes to the nests to gather eggs. Small traditional eight-man[4] boats with outboard motors ferry the hunters from Kirkja for the 30 minute trip to the steep cliffs at Norðberg. The boats are manoeuvred by oar close to the cliffs where the hunters jump from the boat, with socks worn on the outside of their shoes for greater grip, on to the slippery rocks as the waves crash on the steep cliffs that rise up from the sea and are lost in the thick fog above. The puffins are caught on long poles with a Y-shape frame at one end where a small nylon net is fixed. The men, often secured with ropes, climb steep cliffs where they scoop puffins out of the sky as they return to their nests from the nearby ocean feeding grounds. The sky is black with hundreds of thousands of the small birds that dot the sky like shrapnel, and which circle together anti-clockwise as they come in and go out to sea to catch herring.

When a puffin with a herring in its mouth is caught in the hunter's net it is presumed to be a breeder and is released from the net unharmed. A puffin with nothing in its beak, however, is presumed to be a two-year old, yet to breed, and is quickly dispatched by having its neck wrung. The dead puffins are packed into sacks to be transported back to the village by boat. The next few days on the islands are spent plucking the puffins. The loud vibrating whirr of the mechanical puffin-plucker, with its four rows of twirling rubber teats (which pull the feathers from the birds), could be heard throughout the day and late into the night. The black and white birds with their brightly coloured orange, white, red and blue beaks, that a few moments ago were lying side by side serenely on racks in the shed, like proving loaves of dough, are now stripped bare and lined with tiny rows of pimples where the feathers once were before being dropped on to a pile of other naked birds laying on the floor ready to be gutted.

In the next room of the shed, on a small stainless steel bench with an inset tub, the puffins have their feet, head and wings removed before having their innards pulled out. On the opposite side of the room is another bench stacked high with large cod, which receive similar treatment before being filleted and added to the growing stacks of puffins and fish, which are divided among the family groups to be consumed during the winter months.

While haymaking was the major activity of village life on the islands over the summer, it is also the season when pilot whales are most often sighted. My first experience of the *grindadráp* occurred unexpectedly, as *grind* (a school of pilot whales) always do, after one long day of turning the hay and finally tying it down for the night, as we were walking back through the village of Leirvík. An excited neighbour called to us that a *grind* had been sighted not long ago north of Mýlingi and was reported as being about four nautical miles off the coast of Streymoy, the neighbouring island and whaling district. We were quickly in the back of a small truck, which made a couple of stops to pick up other men who would also join the hunt, either as crew on one of the numerous boats participating in the drive, or alternately, to be ready with a whaling knife on the shore of the whaling bay at either Trørnuvík or Norðskála to kill the whales when they were beached. Norðskála was a whaling bay that had not received a

---

[4] Traditional Faroese boats are categorised by the number of men that row them.

*grind* for almost 20 years, and this seemed to increase the excitement as messages were quickly relayed from mobile phone to mobile phone.

Soon I was involved in the hunt as part of a crew on one of the many boats involved in driving the whales. During the drive, the school of pilot whales split up, with many of them escaping back out to sea. The remaining 31 whales were driven ashore at 10:45 pm and swiftly killed. The following morning, when the whalers were being allocated their shares, I was down on the wharf and shown how to flense a pilot whale in the traditional Faroese manner, and how the shares were divided between the hunters.

The second time I was involved in a *grindadráp* was in August 2006, when word of a *grind* also arrived unexpectedly. It was the late afternoon, and I had only arrived back from Eysturoy where the whales had been spotted a few minutes earlier, when I received a phone call telling me a *grind* had been sighted and it was being driven toward the whaling bay in Gøta. This time I joined scientists from the Faroese Museum of Natural History who attend the majority of *grindadráp* to collect biological samples for an ongoing marine mammal research program. I arrived just as the killing of the 131 pilot whales was concluding. The killing was well-organised and very swift, taking 10 minutes from the beginning to the end of the slaughter. The whales were then towed to the nearest wharf where they were lifted by crane and laid out on the wharf to be opened up and then assessed. My role this time was assisting the scientists take samples from the pilot whales and weighing some of the eight bottle-nosed dolphins (*Tursiops truncatus*) that were mixed in with the pilot whales and also taken in the hunt. I arrived back in Tórshavn at 3:30 am the following day and was wakened at 7:30 am with the message that a school of dolphins had been sighted in the fjord between Eysturoy and Kalsoy and were being driven toward the whaling bay at Klaksvík. Again, I accompanied the scientists from the museum to the whaling bay. We arrived as the killing was concluding. Four hundred Atlantic white-sided dolphins (*Lagenorhyncus acutus*) had been driven ashore and another 300 or so were being driven back out to sea as the whaling foremen had made a decision that the whaling district would have enough meat and blubber. Here, helping the scientists, I was assigned the task of sexing, and measuring body-length and the thickness of the blubber of the 52 dolphins being sampled. The other members of the team collected the stomachs, testes, ovaries and livers, which would be analysed during the winter period.

Autumn approaches in the Faroes with a rapid shortening of daylight and a noticeable decline in the zenith of the sun. Autumn is the season when the sheep come down by their own will from the summer pasture in the mountainous outfields. It is also the time when households kill some sheep and prepare them for winter food. Each October, every family that raises sheep kills a small number of them over a period of about two weeks. Even in the capital Tórshavn one notices sheep, either tied up to clotheslines or fences waiting their turn for slaughter. The sheep are slaughtered, often in a room under the house where they are skinned and butchered by hand. During this time, much of the food eaten in the villages is from the sheep including the liver, kidneys, the blood—made in to very sweet blood puddings with raisins—and the sheep heads which have the wool burnt off and always after at least a week's wait for the meat on the head to mature, are then boiled and eaten. The potatoes are dug and stored and the rhubarb is harvested and made into jams. Hares, an introduced species, are hunted during a short season and, like much of the hunted food in the Faroes, are hung outside for several days so that the meat can mature. Cormorants and black guillemot are also harvested. Many of the small boats are hauled up and stored away in neat rows of brightly painted boat sheds that line the many small harbours dotted around the fjords, safe from the winter storms, where they are dried and repaired in preparation for the next summer. Autumn also sees the departure of the millions of sea birds, each leaving on a pre-determined day marked on local calendars, and silence comes to many parts of the islands. The first snows often arrive at this time, dusting the tops of the hills and adding a chill to the wind.

The winter months are the major time of storms, when Arctic winds and huge seas often lash the islands. In winter the sun is only up for five hours a day and the winter solstice is marked with the Christmas festive season when many expatriate family members return home, as do seamen and fishermen from extended voyages. Weekly ferries to Norway and Shetland cease to operate, as do a number of international air routes. It is the time when the sheep are in the infields close to settlements, or are housed indoors, where they are hand-fed silage. December sees the slaughter of many geese for the traditional Faroese Christmas dinner. At this time, many houses in Tórshavn appear to have an influx of geese in their yards or nearby fields. A short while after their appearance they are slaughtered and hung to wire fences to mature for a few days before being eaten. December also sees some of the best cod fishing close to shore. Potato beds are prepared in February and March for planting in spring. Although winter is a time of few pilot whale drives, those caught in winter are considered superior to summer catches. This is because the blubber (*spik*) on the pilot whale (a Faroese delicacy) is thicker in winter after a good summer feeding.

During the period of rough winter weather, many of the fishermen are home from sea with time on their hands, providing the opportunity for the *Grindamannafelagið* (Faroese Pilot Whalers Association) to hold many of its regional meetings in the various whaling districts. This time of year provided me with the opportunity to attend as many of the meetings as I could to listen to what was being discussed about the pilot whale hunt, new and ongoing research proposals, the role of the whalers in the research, and new technologies that were being tried. The meetings also provided me the opportunity to meet many whalers from the smaller villages around the islands, and to ask questions about the management of pilot whaling occurring in the different whaling districts.

With the arrival of spring, the sun returns and the winter shadows cast by the mountains slowly recede. Lambs are born and seabirds, in the millions, return. Again, each species' arrival is marked on the local calendars, their arrival signalled by their shrills and squawks. Once they have nested and laid eggs, the Faroese begin the collection of the eggs in May for a short period. Men scale the steep cliffs on ropes and gather eggs from the nests. Potatoes are planted and halibut are the target of local fishing trips. The abatement of the storms, long hours of daylight and a calmer sea makes it easier to spot schools of pilot whales as they swim close to the islands.

As can be seen from this fieldwork narrative, much of my time in the Faroes involved participating in a variety of community-based resource management activities, specifically those connected with the provision of subsistence food to the households. To sharpen the focus of my fieldwork, and to provide complementary sources of data I chose to adopt a case study approach, as this allowed me to understand the complex social phenomena between the Faroese and the local resources they utilise for food. A case study approach, as well as providing for the inclusion of varied sources of data, also provides for various methods to be employed in data collection relating to Faroese whaling.

**Case Study Research**

The case study approach has been defined as "research [that] involves an in-depth examination of a few people, organisations or groups over time" (Lawrence-Neuman 1994:27). This study of Faroese pilot whaling and its management institutions utilizes a case-study research approach because it "allows an investigation to retain the holistic and meaningful characteristics of real-life-events" (Yin 1984:14). In addition, it is used because, as discussed below, the institution of Faroese pilot whaling is a complex social phenomenon, and as Yin (*op. cit*) informs us: "the distinctive need for case studies arises out of the desire to understand complex social phenomena." Such an approach is important when examining Faroese pilot whaling, as the activity of hunting pilot whales cannot be examined in isolation, outside of the context of Faroese culture and society. Pilot whaling is a real life event, which contributes in significant fashion to families' and communities' food security needs now, as in the past, and is best understood within its wider historic, cultural and economic context.

A major strength of case study research is that it provides the opportunity to use various sources of data from various research methods that can be considered together to provide a rich description of a community and its institutions over time. Case study sources and methods are varied (*ibid*: 78-88) and include:

- documentation;
- archival records;
- interviews;
- direct observation;
- participant observation; and,
- physical artefacts.

The case study approach provides the opportunity for the Faroese pilot whalers to be involved in the research itself by providing them with opportunities to ask the researcher questions about the research, or by pointing out issues and contradictions as they see them. This is important, as noted in an essay on research by Paulo Freire, who wrote: "If I perceive the reality as the dialectical relationship between subject and object, then I have to use methods for investigation which involve the people of the area being studied as researchers. They should take part in the investigation themselves and not serve as passive objects of the study" (Freire 1982:30). Such an approach in this context is especially important, particularly in regard to the critique of Jakob Jakobsen (1904) that introduced this chapter:

> Existing popular descriptions of the Faroes and their people lack a cultural-historical perspective. It is clear that it is foreign eyes that have observed, and foreign pens that have written. Outward activities are described, but without insight into the inner life of the population, without insight into the features peculiar to the culture (Jakob Jakobsen 1904, translation cited in Sanderson 1992:67).

For this reason, it was important that I was actively involved in Faroese subsistence and to have an open dialogue with those engaged in pilot whaling and the wider management of pilot whales in the northeast Atlantic, in order to gain insight into the inner life of the population and characteristics peculiar to Faroese culture.

***Documentation and Archival Records.*** There is a rich variety and vast amount of data available through analysis of documents and archival records on Faroese pilot whaling. This is because the Faroese pilot whale hunt is one of the oldest and most continuously documented wildlife hunts on record, with catch records dating back to the $16^{th}$ century (Bloch 1994). Such documents and archival records provide the opportunity for commentators, and, to some degree, managers of *grindadráp* in the past, to have a voice in a present-day analysis of Faroese pilot whaling. Documentary research and analysis of the historical and contemporary biological, ethnographic and legal papers and archival records provide detailed information on Faroese pilot whaling and its management institutions.

While much of the historic and legal material on Faroese pilot whaling is written in Faroese, I obtained translations of the most important documents needed to situate community-based management of pilot whaling within an appropriate historical, cultural, political, and economic framework.

***Interviews.*** Interviews provide another rich body of data on pilot whaling and its management from the perspective of numerous people involved in various stages of the hunt and its management. Interviews are important as they provide an emic, or 'insiders',' view of events rooted in history and culture. During fieldwork, key people associated with various local institutions related to the conduct of the pilot whale hunt were interviewed; these individuals included biologists, veterinarians, government officials, pilot whaling foremen, hunters (including those crewing the boats and those on shore undertaking the slaughter) and consumers of the products of the hunt. The interviews were structured or unstructured. In structured interviews, key people were asked the same questions to ascertain how various people with differing roles and responsibilities in the hunt and management viewed the same issue. All the interviews used in this book, except that of Schønning Eysturoy, were audio recorded during the fieldwork and later transcribed. The Eysturoy interview was undertaken in 1996 and a transcript provided to me when I was seeking information on technology used in the hunt. Real names are used in this book for all those interviewed; a list is supplied at the end of the bibliography. The quotations from interviews were selected because many of them demonstrate an insider's view of Faroese whaling.

Other interviews were very informal, following an unstructured format, and often involved asking a series of questions at the conclusion of a meeting, or as we undertook various tasks like hay making, hunting, gathering seafood, driving from village to village, or discussions over a meal.

***Direct Observation.*** Yin (1984: 85) notes that "by making a field visit to a case study 'site' an investigator is creating the opportunity for direct observations where relevant behaviours or environmental conditions will be available for observation." Direct observation is useful as it assists with the formation of certain questions, *e.g.*, when an activity was observed to be carried out differently in one whaling district compared to another, then people were asked the reasons for such differences.

Another example of direct observation can be seen in my attendance at meetings of the *Grindamannafelagið* and in other activities, including the pilot whale hunt, the division and distribution of the products of the hunt, the protocols of flensing, the sharing of the catch, all events where customary behaviour can be seen taking place. Direct observations made from boats used in the pilot whale drive also provided opportunities to examine social organisation, technology, and environmental conditions.

***Participant Observation.*** Participant observation was one of the main research techniques employed during fieldwork in the Faroes. Yin (*ibid*: 86) describes participant observation as "a special mode of observation in which the investigator is not merely a passive observer. Instead, the investigator may take a variety of roles within a case study situation and may actually participate in the events being studied."

As discussed above, in the narrative describing my fieldwork, participant observation allowed me to undertake various roles while in the Faroes, and in so doing allowed me to actively participate in Faroese life and thereby gain a more informed understanding of why certain activities were being undertaken. This was probably most notable at the end of the pilot whale hunts. On one occasion, when there were many tourists present (a large cruise ship had recently berthed in Tórshavn) during the assessment of the beached whale carcasses, it was quite noticeable how Faroese people whom I did not know engaged with tourists, as opposed to how they engaged with me while I was busy taking the weights of the bottle-nosed dolphins. Nobody changed their behaviour. Bernard (1988:342) comments on this when noting that participant observation is beneficial to field work as it reduces the problem of reactivity—that is, people changing their behaviour when they know that they are being studied.

Participant observation gives the researcher an intuitive understanding of what is going on in a society, and allows him or her to speak with confidence about the meaning of data and make strong statements about cultural facts that have been collected. It extends both the internal and the external validity of what the researcher has learnt from interviewing and watching people. In short, participant observation helps the researcher understand the meaning of their observations (Bernard 1988:344).

However, some major problems related to participant observation may occur, and include potential biases that may be produced. For example, when "the investigator has less ability to work as an external observer and may, at times, have to assume positions or advocacy roles contrary to the interests of good scientific practices" (Yin 1984:87). It is also possible that a researcher may omit data in order to confirm their own pre-established beliefs, leaving the method open to the criticism of bias. With this in mind it is important to note, however, that the issue surrounding data production as mediated by the researcher is not peculiar to participant observation.

As mentioned earlier, I have been involved in an advocacy role for other whaling communities, and because of this I am open to a charge of possible bias. While these criticisms are real, the trade-offs between the opportunities and the problems requires serious consideration. In relation to my research on Faroese pilot whaling, the opportunity to gain an understanding and to perceive everyday life from the viewpoint of someone from 'inside' the Faroese pilot whale hunt and its management institutions, greatly outweigh the problems of perceived bias. From that perspective, the opportunities provided to participate and to observe various events within Faroese social life to create knowledge and understanding of complex social institutions of resource use, in my view, far outweigh any associated potential problems of bias.

***Physical Artefacts/Material Culture.*** Physical artefacts, or objects that are sometimes referred to by the broader term 'material culture,' are important sources of data especially when undertaking a study of resource use. Physical artefacts employed in resource use can provide clues to the sustainability of a community's practices over time and provide evidence of the evolution of technology and it uses. Physical artefacts can either be collected, or observed as part of fieldwork. They have been used extensively in anthropological research; however, in the past they have often been removed from the cultural context and interpreted in isolation from the activity they were employed to facilitate. Material culture is not just the physical objects but also includes the skills, knowledge and behaviour that surround an object or activity. For example:

> Though we tend to think of economic and technologic adaptation as being separate realities, though admittedly linked, the relationship is indeed much more profound and intimate... [t]echnology should be considered as artifice, not artefact, that is the behavioural act rather than the technical means of accomplishing a particular end. In this sense the emphasis becomes placed on the social and ideological, rather than the merely material aspect of the activity (Freeman 1990:7).

Freeman provides the following example to illustrate this important point:

> [C]onsider Inuit winter seal hunting. This is ordinarily carried out at a seal's breathing hole on the sea-ice... . It appears that there

> are almost thirty separate implements included in a seal hunter's winter tool kit, from the toggle harpoon, to the harpoon rest placed on the snow, to the probe used to determine the exact shape of the passage through the ice the seal will use to reach the surface to breathe, to the different types of indicator placed at the top of this passage to signal the presence of the seal, to the polar-bear skin mat the hunter stands on noiselessly to await the seal, and so on. It is true that all this equipment was used, but no Inuit father would feel he had equipped his son for a life of seal hunting if all he did was instruct his son in how to manufacture such an elaborate tool kit. What the hunter needs above all else is patience, together with an understanding of seal behaviour and of environmental phenomena, an appreciation (sometimes supernatural) why seals sometimes do not offer themselves to the hungry hunter and his family. To be a seal hunter requires hours of learning over several years about much more than the technical aspects of tool manufacturing (*op. cit*).

As in the case of the Inuit hunter, where the passing of knowledge between generations is just as, if not more, important as the physical artefacts used in the hunt, so too it is for the Faroese. If knowledge is not passed down through the generations, along with the technology, the success of the pilot whaling hunt and the procurement of food is threatened.

In relation to pilot whales as a flow resource, the North Atlantic Marine Mammal Commission (NAMMCO) has established that an annual catch of 2,000 pilot whales by the Faroese corresponds to an exploitation rate of 0.26 percent of the present best estimate in abundance of pilot whales (which currently stands at 778,000 animals). A catch rate of little more than one-quarter of one percent of the total population indicates that the flow of the North Atlantic long finned pilot whale resource is sustainable.

A brief narrative of the author's fieldwork experience in Faroese villages clearly demonstrated that Faroese society corresponds to what Bulliet (2005) refers to as a 'domestic society,' one where Faroese people living in small settlements are heavily reliant on wildlife for food and consequently accept the need to kill animals to ensure food security and experience few moral qualms in consuming animal products.

Research methods employed in this study were appropriate to enable an 'outside researcher' to develop necessary social relationships with the Faroese people in order to gain insights into the daily life of the population and the common property resource institution associated with the utilisation of whales. The case study approach used in this research provided the opportunity to obtain a variety of data that provide a rich description of Faroese society and its pilot whaling institution.

# *Theoretical Framework*

> *In the absence of any substantial economic or nutritional need, the Faroese have increasingly turned to their 'history' and 'culture' to provide justification for the continuation of the pilot whale hunt. This, given their rather successful integration into world market economies, brings them to a rather contradictory and paradoxical position. On the one hand, they claim to be an isolated, self-sufficient and traditional society, while on the other, their standard of living depends on them being just the opposite. The Faroes cannot have the best of both worlds. Cultures, just like economies, are susceptible and, indeed, are required, to change. This cannot be more clearly illustrated than by the less-than-traditional features that have come to characterise the modern pilot whale drive. High-powered boats and modern communications equipment have come to replace the traditional methods of the hunt* (Whale and Dolphin Conservation Society 2001).

Over the last 20 years, research on common property resources, or what are sometimes called common-pool resources, and local self-organised governance regimes of common property resources has rapidly developed. The establishment in the 1980s of the Panel on Common Property Resources at the U.S. National Academy of Sciences was seen as an important catalyst in developing a sophisticated understanding of common property resources and associated management institutions (Ostrom and Schlager 1996). Subsequent publications presented case studies detailing systems of rules, rights and responsibilities that guide and control human use of natural resources held in common, many of which are dynamic adaptive systems that have ensured sustainability of both the resource and the resource users over long time periods (*e.g.*, Berkes 1989; Berkes *et al.* 1989; Bromley 1989, 1991; Pinkerton 1989; Fenney *et al.* 1990; McCoy and Acheson 1987; North 1990; Ostrom 1990, 1992; Dasgupta and Mäler 1992; Berkes and Fölke 1998; Hanna *et al.* 1999).

At the conclusion of this discussion of common property resources, the use of pilot whales and the institution of pilot whaling on the Faroes is examined in relation to what Ostrom (1990:67-68) has termed 'design principles' that have been identified in the common property resource literature as being essential features for successful common property resource management outcomes. A successful common property resource institution is one that sustains both the resource and the local management institution over time. This chapter concludes by identifying a number of common property resource institution design principles that can be used to determine the robustness and sustainability of both the *grindadráp* and the North Atlantic long-finned pilot whale population.

This chapter provides definitions of the terminology and concepts used in this analysis of Faroese whaling within a common property resource institution that respond to ambiguities in the use of terms used by animal rights and environmental organisations in their criticism of Faroese pilot whaling. Bromley also draws our attention to this concern within the literature on natural resources:

> In the literature on natural resources and environmental policy, it would be difficult to find the idea (that is the concept) as misunderstood as 'commons' or 'common property.' The mischief that arises from the continuing failure to understand common property is perverse in both scholarly discussions and public policy formation. On the former front, scholars will show no hesitancy to expound on the problems inherent in common property without the benefit of first defining 'property,' and without betraying any understanding of the historical and contemporary facts surrounding common property regimes (Bromley 1992:3).

## The 'Tragedy of the Commons'

Such an example of conceptual misunderstanding can be seen in Garrett Hardin's seminal essay entitled *The Tragedy of the Commons* (Hardin 1968). The tragedy referred to was his postulated widespread over-use of a resource held in common, illustrated by a hypothetical example of medieval English herders, whose access to pasture (the exploited resource) was unlimited and occurred in the apparent absence of rule-making to restrain over-use of the resource. According to Hardin, the herders could bring any number of cattle to graze because the pasture, described as a 'commons,' was open to all. Because the right to graze was unspecified and unlimited, herders, acting individually, continued to add more and more cattle, taking only their own benefits and costs into account while ignoring the collective effect of their individual actions. Because there was no system of rules that described how grazing was to occur, the pasture soon become overgrazed and all the herders suffered. Hardin argued that the eventual fate of all resources held 'in common' is over-exploitation because access is unrestricted and there is no incentive among individuals favouring collective resource protection.

Hardin's conclusion that "freedom in the commons brings ruin to all" (Hardin 1968:1244) was "accorded by some the status of scientific law" (Berkes 1996:89). This was because Hardin portrayed 'the commons' as a model where people seemed helpless and unable to co-operate for the greater good of the community, leading many to the conclusion he advanced that management of common property resources should be removed from local communities and either privatised or controlled by a state authority to ensure optimum use. However, Hardin had made a serious mistake in confusing common property resources with open access to resources.

Social scientists working with indigenous peoples and local communities in many different regions of the world saw evidence in the operation of a wide range of common property regimes that Hardin's thesis would not stand up to historical and ethnographic scrutiny, and that his generalisation about 'the commons' was inappropriate (Dasgupta 1982; Cox 1985; McCay and Acheson

1987). Many asserted that Hardin's thesis did not critically identify the type of property regime susceptible to such a process, arguing that his critique did not apply to common property, but rather, to open-access regimes (Fenny *et al.* 1990) and "although open access is a distinctive 'commons' problem, it is not definitive of common property" (McCay 1996:113).

A continuing failure to understand common property is mirrored in imperfect understandings of other concepts and terms related to natural resource management, including 'subsistence,' 'sustainability,' 'culture,' 'institutions,' and 'institutional performance.' This confusion may result from the narrowly unidisciplinary training of natural resource specialists in a variety of different disciplines, or it may have more to do with political or ideological motivations of commentators resulting in their using concepts and terms without providing accepted definitions of the terms used, or engaging in what has been termed 'trans-science' (Weinberg 1976; Freeman 1994). We have seen earlier how the mischievous misuse of the term 'whale' has led to the wide acceptance of the mythical 'super-whale' as the symbol for all species of cetaceans. To avoid such a situation, this chapter provides an overview of the theoretical framework and related concepts used within this current analysis of Faroese whaling.

## Culture

The use of the term 'culture,' as it has been applied to Faroese pilot whaling by some environmental and animal rights' organisations, is first examined. As mentioned above, the Whale and Dolphin Conservation Society's (WDCS) campaigns against Faroese whaling assert that the Faroese could not continue the subsistence pilot whale hunt while at the same time engaging in world market economies. The WDCS stated that: "The Faroes cannot have the best of both worlds. Cultures, just like economies, are susceptible and, indeed, are required, to change" (WDCS 2001). This assertion is problematic in its belief that culture and economy can be separated as if they bear no relationship one to the other. Culture and cultural differences are important in understanding resource use and the function of property rights regimes. Thus in this current analysis of Faroese whaling, culture is not treated as an isolate, nor as a source of explanation in itself, but as a variable that is linked to political and economic forces, social institutions and biological processes.

In this context, the notion of culture is similar to that advanced by South African anthropologist Adam Kuper who summarises his understanding of culture in the following way:

> [T]he more one considers the best modern work on culture by anthropologists, the more advisable it must appear to avoid the hyper-referential word altogether, and to talk more precisely of knowledge, or belief, or art, or technology, or tradition, or even the concept of ideology (though similar problems are raised by that multivalent concept). There are fundamental epistemological problems, and these cannot be solved by tiptoeing around the notion of culture, or refining definitions. The difficulties become most acute when (after all the protestations to the contrary have

> been made), culture shifts from something to be described, interpreted, or even perhaps explained, and is treated as a source of explanation in itself. This is not to deny that some form of cultural explanation may be useful, in its place, but appeals to culture can offer only a partial explanation of why people think and behave as they do, and of what causes them to alter their ways. ***Political and economic forces, social institutions, and biological processes cannot be wished away, or assimilated to systems of knowledge and belief*** (Kuper 1999:x-xi, emphasis added).

It is for this reason that my analysis of Faroese whaling will focus on the political and economic forces, social institutions, and biological processes that have shaped and continue to shape the Faroese use of whales. In doing this I propose, following Kuper (*op. cit*), to move away from viewing culture as something to be treated as a source of explanation in itself, to one where culture remains something to be described, interpreted or explained and that can only be understood in terms of the political and economic forces, social institutions, and biological processes in which culture rests. In this regard, as I will demonstrate, contrary to the allegation made by the WDCS, the Faroese can and do have 'the best of both worlds,' as do all societies engaging in both subsistence and global market economies (referenced in, *e.g.,* Langdon 1986; Headland and Reid 1989; Peterson and Matsuyama 1991), including marine mammal hunters in the North Atlantic region, where:

> It is erroneous to consider marine mammal hunting as an activity of bygone days. On the contrary, hunters on both sides of the North Atlantic have insisted on their rights to carry on the trades of their ancestors within the framework of the modern world. This does not mean that the societies are static or unwilling to change. They are continually adapting to changes in the natural as well as the social environment (Kalland *et al.* 2005:56).

## Theoretical Framework

The first theoretical concept I introduce is that of political economy and its related world-systems theory, introduced here for reasons that Marcus and Fisher explain:

> The usual objection to interpretive ethnography is that 'cold,' 'hard' issues of power, interests, economics, and historical change are elided in favour of simply portraying the native point of view as richly as possible. While this objection has had some validity, many interpretative ethnographies now try to take account of power relations and history within the context of their subjects' lives. However, there seems to us to be a more radical challenge in this, by now conventional charge against 'symbol and meanings' ethnography: how to represent the embedding of richly

> described local cultural worlds in larger impersonal systems of political economy. This would not be such a problematic task if the local cultural unit was portrayed, as it usually has been in ethnography, as an isolate without forces of market and state impinging on it. What makes representation challenging and a focus of experimentation is the perception that the 'outside forces' in fact are an integral part of the construction and constitution of the 'inside,' the cultural unit itself, and must be so registered (Marcus and Fisher 1999:77).

A discussion of political economy and world-systems theory will situate Faroese whaling within a broad historic economic and political framework, which will assist in addressing the undercurrent portrayed in environmental and animal rights organisations' campaign material which maintains that the Faroes is a discrete isolate with its own antiquated and outdated resource-use practices, disconnected and far removed from the influences of global political and economic forces. Such a view is evident in statements such as, "if they want to kill whales in the traditional way, that's fine by us, if nothing else about their way of life, significantly anyway, has changed" (WDCS 1994).

**Political Economy**

While the Faroes may be geographically isolated, the Faroese do not live in political and economic isolation from the world economy. As discussed below, outside forces are an integral part of the construction and constitution of the 'inside' of the cultural unit itself. In this regard, the Faroese have a long and complex history of interaction with global political and economic forces. The islands' 1,000-year history of connectivity with these outside forces, primarily through relations with Europe, is important in understanding Faroese institutions, such as the pilot whale hunt, today. Therefore, it is important to examine Faroese institutions, not in isolation, but through careful historical analysis of dynamic relationships between differing societies with differing modes of production. As is now generally recognised:

> Most local cultures worldwide are products of a history of appropriations, resistances, and accommodations. The task for this subtrend in the current experimental moment is thus to revise conventions of ethnographic description away from a measuring of change against some self-contained, homogenous, and largely ahistoric framing of the cultural unit toward a view of cultural situations *always* in flux, in a perpetual historically sensitive state of resistance and accommodation to broader processes of influence that are as much inside as outside the local context (Marcus and Fisher 1999:78).

This relationship between the inside and outside is best viewed through the concept of political economy. Political economy is an old term for the study of economics, which has been inseparably bound up with the study of history, politics and statecraft. The use of, and reference to, the term declined during the

19th century and as a result, the study of economics became isolated from the study of politics. However, in recent times, there has been a resurgence in the use of the term political economy (Marcus and Fisher 1999:79).

Currently, there are three major references to political economy in contemporary usage, and in this research context I use political economy to mean a "more generically defined interest in the mutual determination of political processes and economic activity in a historically viewed world system of nation states" (*op. cit*).

**World-System Theory**

World-system theory "has powerfully communicated the idea that the significance of any particular project of research in history or ethnography lay in its placement within a larger world-historic framework of political economy" (*ibid*: 81). Immanuel Wallerstein introduced world-system theory in the early 1970s when he:

> directly challenged the failure of development theories of the 1950s and 1960s and proposed within the ahistoric and separated disciplines of political science, economics, and sociology to explain what was happening in the developing world. The contemporary third world, or any other part of the globe, had to be understood in the context of the history of a capitalist world economy developing since the 16th century. A theoretically grounded account of this history was to be Wallerstein's coordinating vision for interdisciplinary research. Wallerstein thus insisted that the only effective social theory was one that was tied to detailed considerations of world-historical events and processes (*ibid*: 80).

In this way:

> The world-system perspective is a macroview of society and history, but its very attraction has been its simple (and at times, simplistic) theoretical formulations, contrasted with its emphasis on working out its concepts through the interpretation of historical detail. It has thus served less as a full-blown theory than a framework for debates and discussions. And these debates depend vitally on ethnographically and historically sensitive research in political economy...
>
> [T]he world-system theory survives today primarily as a general orientation that thrives on detailed studies of regions and historical periods. Rather than their emphasis being on the 'system,' appropriate for the times, political economists have focused their attention on close analyses of the historic and ethnographic conditions of regions and locales (*ibid*: 80-81).

World-system analysis "makes central the study of long-term, large-scale social change" (Wallerstein 1991:4); it also embraces political economy's traditional emphasis on core-periphery relations and the role of the state and the question of who bears the cost of development. Wallerstein rested his account of the capitalist world-system on distinctions between core, semiperipheral, and peripheral areas of world political and economic development, and on the historic conditions for shifts in the relationships of these areas (*ibid*: 4-6).

This can be seen in the European colonial order, which focused on the extraction of material wealth from colonised regions (the periphery) to their colonial power's heartland (the core). This process of colonisation economically linked together communities and territories that, in many cases, had led a fairly autonomous existence before colonisation.

This new economic order did not spring up painlessly by itself. It was imposed and maintained by force, and for this reason many anthropologists describe the colonial order as a political economy, which is:

> a holistic term that emphasises the centrality of material interest and the use of power to protect and enhance that interest. The colonial political economy created three kinds of links: connecting conquered communities with one another within a conquered territory; different conquered territories with one another; and all conquered territories with the country of the coloniser (Schultz and Lavenda 2001:4).

Major large-scale social processes, such as incorporation into the capitalist world economy, are not abrupt phenomena, for:

> [t]hey emerge from the flow of ongoing continuous activities. While we may give them dates retrospectively and approximately, the turning points are seldom sharp and the qualitative changes they incarnate are complex and composite. Nevertheless, they are real in their impact and eventually they are perceived to have occurred (Wallerstein 1989:32).

Wallerstein proposes that with the expansion of the capitalist world economy "there resulted a greater concentration of the most profitable economic activities within particular state boundaries" (*ibid*: 33). What a region could produce in terms of tradable goods would determine its place within a broader colonial economic order. Those that produced few tradable goods, or were not geographically strategic in maintaining the broader colonial economic order, found themselves on the periphery.

According to Wallerstein, world-system analysis has three defining characteristics. First and most obvious, the appropriate 'unit of analysis' for the study of social and/or societal behaviour is a 'world-system.' This approach highlights dynamics between local, national and global spheres and denies that the 'nation state' can be viewed as a relatively autonomous 'society' that 'develops' over time (Wallerstein 1991).

Second, world systems are explicitly historical and dynamic; they have 'beginnings, lives and ends.' This illuminates the transitory nature of institutions and the fact that there are uncertainties at periods between one historical system and another. "This stance," says Wallerstein "makes clear that structures are not 'immobile.' It insists, in addition, that there are 'transitions' from one historical system to its successor or successors. It is this pair, the space of a 'world' and the time of a 'long duration,' that combine to form any particular historical world-system" (*ibid*: 267).

The third characteristic of the world-system is the capitalist world economy. Wallerstein (*op. cit*) lists a number of characteristics, which he describes as being components of the capitalist world economy, including, *inter alia*:

- the ceaseless accumulation of capital as its driving force;
- a structural division of labour between core and peripheral zones;
- the large and continuing role of non-wage labour alongside of wage labour;
- the non-primordial character of states, ethnic groups, and households, all of which are constantly created and re-created;
- the view that the capitalist world economy began in Europe and later expanded to the entire globe;
- processes whereby antisystemic movements simultaneously undermine and reinforce the system; and
- cyclic rhythms and secular trends contributing to inherent contradictions of the system.

World-systems analysis also recognises variations within the core and the peripheral zones. Eric Wolf (1997:297) acknowledges the diversity of relationships within the world economy when he draws a distinction between the capitalist mode of production and 'the capitalist world market':

> The capitalist mode of production may be dominant within the system of capitalist market relations, but it does not transform all peoples of the world into industrial producers of surplus value... rather it opens up the question of how that capitalist mode relates to other modes of production...[and] allows us to take note of the heterogeneity of different societies and subsocieties making up the system rather than obliterating the heterogeneity in the dichotomies such as 'core-periphery' or 'metropolis-satellite' (*op. cit*).

World-systems analysis also recognises that societies on the periphery of politically and economically strong cores can, to some degree through local decision-making, shape their own future in a positive manner. This is an important feature of world-systems analysis, as it does not treat those communities on the periphery of the world economic order as being passive players with no ability to actively engage with a much broader political and economic order to determine, to some degree at least, their own futures.

**Subsistence Modes of Production**

While world-system theory is the most appropriate way to examine Faroese relationships with a wider economic order, it is also important to first understand resource use and mode of production at a local level through a community's own socio-economic system. A greater understanding of subsistence modes of production comes from an analysis of the patterns of use of wild and renewable resources, which can be understood only in relation to the 'socio-economic systems' of the communities within which they occur. In this regard, R.J. Wolfe informs us that "a 'system' is a set of interacting, interrelated, or interdependent elements forming a collective entity," with a socio-economic system being "a functionally related set of elements which provides material and social support for a community or regional population" (Wolfe 1983:248). It is the socio-economic system which comprises the basic structural relationships underlying the material and social well-being of a group of people. A breakdown in the socio-economic system can potentially lead to negative social indicators, such as economic hardship, community breakdown through either the loss of social institutions or increased migration, low birth rates, high unemployment and inflation rates (*ibid*: 248-249).

Wolfe deconstructs the socio-economic system into several interrelated elements, which will assist our understanding of the Faroese use of pilot whales. The first of these are socially-constituted groups, such as family units, economic firms and corporate organisations. These groups are organised to perform essential activities for a community, including food and material production, exchange of goods and services, education, and the rearing of children. Within these activities, a division of labour is frequently provided in learned social roles, such as occupations and job tasks (*ibid*: 249). Additionally, as noted earlier by Wallerstein (1991:267), there is the importance of the large and continuing role of non-wage labour alongside wage labour. The social groups and the social roles organise human interaction in the system. The second and third elements in the socio-economic system are the mode of production and the economic resource base. The mode of production consists of the technological means for producing, distributing and consuming goods within the socio-economic system. Production technology is used to extract and convert material from the natural resource base of a community into consumable goods. A community's natural resource base includes its land, waters, and their physical and living assets, including whales and fish that are or may be exploited in order to provide a livelihood for the community.

In his text, *Europe and the People without History*, Eric Wolf (1997) characterises a mode of production as a specific, historically-occurring set of social relations through which labour is deployed to wrest energy from nature by means of tools, skills, organisation, and knowledge. The use of the concept 'mode of production' allows us "to inquire into what happens in the encounters of differently constituted systems of interaction—societies—predicated upon different modes of production" (*ibid*: 76). The concept also aims "at revealing the political economic relationships that underlie, orient, and constrain interaction" as well as shaping divergent systems of ideas (*op. cit*).

Economic theory categorises socio-economic systems into three parts. "When referring to 'labour, capital, and land,' social science theory calls them 'social organisation, technology and environment.' Either way, the socio-economic system comprises an arrangement of these factors in a functioning whole which provides for the material support and continuation of a community" (Wolfe 1983:249-250).

> A socio-economic system organises a community or region, and exists at a higher level of complexity than the individual. Individuals operate within the socio-economic system of a community, becoming part of it by birth or immigration. They learn to enact the social roles within the system, and through their enactment preserve and modify the system. However, a socio-economic system of a community has an existence apart from any individual member. The system has a history that predates and a future that outlasts particular members. Thus, the system is not reducible to individual characteristics of its members (such as age, health, personality, income, ethnic status), although these characteristics under certain qualified circumstances might be used as identifying markers of a particular socio-economic system (*ibid*: 250).

Wolfe (*ibid*) goes on to say that it is useful to try to classify socio-economic systems according to characteristics of their social and economic base. As such, small, dispersed settlements worldwide came to be characterised by the production of food and raw material, such as by plant cultivation, animal husbandry, forestry and fishing and hunting activities. The economic base of such communities is what he calls 'food extractive' in nature. This contrasts with large urban areas worldwide, where the economic base of the community is manufacturing, finance, international trade and defence. A food extractive economy is what Wolfe terms a subsistence-based socio-economic system, and it is within such a subsistence-based socio-economic system that we find communities dependent on customary and traditional use of wildlife resources. Without continued access to wildlife resources extreme disruption to the community's social and economic well-being may occur (*ibid*: 253).

To many people, the term subsistence equates with the absence of cash and a struggling existence. Freeman (1993:245) points out that "[t]o the non-specialist the term subsistence relates in important ways to an individual's economic and material circumstances." This limited view contrasts sharply to that of researchers who, through in-depth fieldwork in fishing and hunting communities, can demonstrate the critical importance of wildlife to hunting communities "by the extent to which that activity is central to reproducing its social relations of production, for example, through the socialisation of children, mutual aid and sharing, and reinforcement of stewardship and use arrangements with respect to land and resources" (Usher 1981:61). In that regard:

> A subsistence society can be understood to be a group of people whose production, use and consumption of local resources occurs in ways that are consistent with traditional patterns maintained by kinship-based social structures. Such societies possess detailed knowledge of their environment, and particularly those resources important to their food-production and ceremonial activities.

> Traditional knowledge, required for harvesting and processing subsistence resources, is transmitted from generation to generation principally by oral means and requires an extended period of learning through experience. Knowledge and experience are most often obtained by the individual maintaining close association with an appropriate member of the local community who is often related by kinship or through another alliance-creating institution (Freeman 1993:245).

Subsistence activities have been defined as possessing six fundamental characteristics. These are (with emphasis added):

> - a *mixed economy*, with mutually supportive market and subsistence sectors;
> - a *domestic mode of production*, where production capital, land and labour are controlled by extended kin-based production units;
> - a stable and complex *seasonal round of production* activities within the community, tied to the seasonal arrival, and variable yields, of fish and game resources;
> - substantial non-commercial *networks for sharing, distributing and exchange* of food and materials;
> - *traditional systems* of land/water use and occupancy; and
> - complex *inter-generational systems of belief*, knowledge and values associated with resource uses, passed on between generations as the cultural and oral traditions and customs of the society (Wolfe 1983:272).

With an understanding of political economy, its related world-systems analysis, and local-level subsistence modes of production, it is necessary to next examine the social relationships between people and the resources they utilise, or what have been termed 'property rights.' An overview of property rights will assist in understanding the political relations between Faroese people involved in the subsistence economy and the manner they utilise the pilot whale resource.

**Property and Property Rights**

Political economy and world systems analysis is closely related to property rights. Property is *not* an object, but is rather a social relationship between people that defines the property holder with respect to something of value (the benefit stream) against all others. As such, property is "a political relation between persons," or in some instances, between states; property rights are "an enforceable claim to some use or benefit held by some persons and not by others" (Macpherson 1978:4). Property is used to mean the rights and obligations of individuals or groups to use a resource base (Bromley 1992). Property is a social instrument, and is held via various property regimes or institutions which are chosen for particular social, cultural and environmental reasons.

Property rights have several functions. First, they delineate the population of legitimate owners or users of a specific resource. Second, they specify the allowable actions of the owner or user and their associated responsibilities so that expectations are consistent and enforcement of the rules is possible and cost-efficient. Third, by setting consistent expectations among users, they reduce uncertainty about other behaviour and build social cohesion among the group of users by reducing the chance of conflict (Hanna 1999).

Property rights include a bundle of entitlements defining owner rights, duties, and responsibilities for the use of the resource, or a claim to a benefit (or income) stream, and a property right is a claim to a benefit stream that some higher body—usually the state—will agree to protect through the assignment of duty to others who may covet, or somehow interfere with, the benefit stream (Bromley 1992).

Property rights to natural resources control far more than use of a resource, for they can also manage conservation and sustainability of both the resource and the ecosystem within which the resource is located. Thus the scope of property rights may be focused on either an animal or plant population or stock, or areas of land and/or water in which a particular population or stock lives. Property rights may also include different types of use. Most systems of property rights, for example, are designed for direct uses, such as the catching of wildlife for food, but they are also slowly being developed to include indirect uses, such as the right of the public to enjoy wildlife through non-consumptive use, such as whale-watching (Hanna 1999).

Property rights systems do not simply define and grant rights to users, rather they establish the rights and responsibilities of system participants *vis-à-vis* each other. Property rights systems form part of a community's law, and as such are core elements that differentiate communities—law being "the central art by which communities and culture are established, maintained and transformed" (White 1985:684). The lack of a system of property rights will surely result in chaos, for:

> [t]he basic functions of managing resources—co-ordinating users, enforcing rules, and adapting to changing environmental conditions—cannot be met without a system of property rights. The way that property rights function in any particular context determines whether the natural and human system will interact in compatible ways or conflicting ways (Hanna 1999:117).

The terms 'property rights' and 'rules' are frequently used interchangeably when referring to uses made of natural resources. The way that Ostrom and Schlager use these terms is to recognise that 'rights' are the product of rules, and consequently are not equivalent to rules. This distinction is important, as a property right carries with it an enforceable authority to undertake particular actions in exercising that property right. Because rights define actions that individuals can take in relation to other individuals as regards some 'thing,' if one individual has a right, someone else has a commensurate duty to observe that right. The duty that an individual owes another defines the actions the individual may, must, or must not take in relation to another and to that other's property (Ostrom and Schlager 1996:130).

The various forms of property rights have, in practice, been applied almost exclusively to environmental goods. When resources are thought of as natural capital, it is easy to see that they create value through the size of standing stock and through the flow of resources from that stock. We are familiar with property rights over the stock and flow of natural capital goods. For example, in marine ecosystems, rights can be assigned to the resources in three ways: on capture (competitive fishing); to the standing stock itself (territories for sedentary species); or to the flow of goods from the standing stock (individual fishing quotas). Property rights to goods from marine ecosystems vary from private/community/state property in near-shore areas, to state property in offshore zones, to open access in international waters. However, in relation to international waters, this is slowly changing, for example, through the development of international agreements in regard to highly-migratory fish species.

For property rights' regimes to be successful they must perform certain functions, such as limiting use, co-ordinating users, and responding to environmental conditions. Attached to these functions are the transaction costs of co-ordination, information gathering, monitoring, and enforcement (Eggertsson 1990). Often those who champion private property regimes as the most effective institutional form for resource management and environmental protection claim transaction costs are low when only one person is required to make a decision. Compared to even well-functioning common property regimes this is true, as they require more time and resources to ensure that the views of all co-owners are canvassed in decision-making. However, "the notion of transaction costs is culturally specific—one person's tedious meeting (a cost) may be another's enjoyable activity (a benefit). That is, in some cultures, such meetings take on great social significance and become the venue for a wide array of useful social interactions" (Bromley 1991:131). With this in mind, we can begin to see the role cultural differences can play in the function of property rights regimes:

> The relationship between culture and property rights becomes most exaggerated under conditions of change. When property rights are changed in such a way that they violate preconceived notions of fair distribution (at least for some significant portion of the population), we can predict a move away from self-enforcement of property rights to more dependence upon community and third-party formal sanctioning. Under these conditions, the costs of negotiating, monitoring, and enforcing new rights (the transaction costs) may undermine whatever efficiency benefits have been anticipated by a change in property rights. Should this be the case, a likely result will be the failure of new property rights (Ensminger 1996:180).

**Types of Property Rights Regimes**

There are various types of property rights regimes. "Differences in form derive from the scope of the rights, the type of rights owner and the privileges and responsibilities of the rights owner" (Hanna 1999:118). Different interest groups often promote a particular form of property rights regime because of its ability to contribute to achieving larger goals, such as community stability, equity or economic efficiency. The goal of sustaining marine biodiversity can be achieved with a variety of forms of property rights. "The key requirement is that the system of property rights fits into the ecological, economic and social context in a way that contributes to maintaining ecological function and diversity" (*op. cit*: 118). Resource managers cannot function effectively unless they know the property rights regime they are dealing with (Bromley 1991:24-25).

Property rights regimes can be ordered loosely along a continuum, where the owner ranges from an individual person, to a small community, to the state, or to no-one in particular (Hanna *et al.* 1996). In reality, many marine and coastal resources are held under regimes that combine the characteristics of two or more types of property rights regimes. The four discussed below are ideal, analytical characterizations; they do not exist in the real world. Rather, resources tend to be held in overlapping combinations of regimes. Strictly speaking, for example, pure communal property systems are always embedded in state property systems and state law, from which they derive their authority.

**State Property**

A state property right *(res publicae)* is one where ownership and control of the natural resources lies with the state. Individuals and groups can make use of the resources, but only within established rules. State property regimes remove the majority of management decisions from the users, and generally convey no long-term expectations of tenure. The controlling state agency has the right to make use and access rules, while users have a duty to observe use and access rules determined by the controlling agency (Hanna 1999). An example of state property is conservation estate, such as national parks, where the land and its resources are owned and controlled by the state. Here, the controlling state agency has the right to make rules concerning use and access, and if the public want to use the conservation estate, they must abide by the rules or face penalties. State property does not always mean that decisions of use and access are made solely by the state agency. For example, in recent years there have been examples where the conservation estate is managed by boards or advisory committees with membership drawn from a wide variety of stakeholder groups, including indigenous and local peoples, environmentalists, biologists or government officials. These emerging management bodies, while reflecting common property arrangements to a certain degree, differ sharply from true common property arrangements, because in the case of state property, it is always the state that has the overriding power to make decisions.

**Private Property**

Private property rights (*res privatae)*, are the most familiar property rights in the Western world, and assign ownership to named individuals, including legal entities such as corporations, guaranteeing to those owners a bundle of rights about access and use. All corporate property is, in fact, private property, administered by a state or non-state group. Private property does not bestow full and absolute power on the owner(s) allowing them to do as they wish. In the case of private land and its associated resources, the owner is under specific statutory obligations regarding what can be done with the land and its resources. A private owner of land, for example, may have a right to use water in the streams, but has a duty not to pollute the streams. Private property is often defined as being transferable, and it confers on the rights-holder the legally and socially sanctioned ability to exclude others, forcing them to go elsewhere (Bromley 1991; McCay 1996; Hanna 1999).

**Common Property**

A common property (or common pool) resource (*res communes*), is owned by an identified group of people who have a right to exclude non-owners, the duty to participate in decisions about use, and the responsibility to act as resource stewards. Common property is also far less transferable than private property (McCay 1996; Hanna 1999). "The property owning groups vary in nature and internal structure across a broad spectrum, but they are social units with definite membership and boundaries, with certain cultural interests, with at least some interaction among members, with some common cultural norms, and often their own endogenous authority system" (Bromley 1991:26).

Individual members of the management group (the co-owners) have both rights and duties with respect to use rates and maintenance of the resource. Common property is in essence 'private' property for the group in the sense it is the group who decides who shall be excluded. Common property is not, however, 'everybody's property' and should not be confused with 'open access.' Common property resources require social co-ordination and cooperation among a defined group of users, for:

> Without co-ordination, individuals may in the aggregate use too much too fast, causing the rate of production to fall. Sharing without collective consumption—the commons situation—requires restraint, which in turn depends on co-ordination among users. Otherwise, individuals continue to consume without regard to the diminishing marginal product of the commons as a whole (Oakerson 1992:47).

Social co-ordination, and the institution a community chooses to undertake this important activity, is one of the most essential elements of a successful common property resource regime, and, as such, this feature clearly demonstrates the point that property is *not* an object but is rather a social relationship between people that defines the property holder with respect to something of value (the benefit stream) against all others (Macpherson 1978).

**Open Access**

Open access (*res nullius)* refers to any resource or object that is ownerless, where ownership or possession is open to any person or group, with that resource becoming owned only at, and after, the time of capture or possession.

This absence of property rights can lead people to 'free ride' and over-exploit a resource. There are two reasons why open resources regimes exist. First, because the natural resources have never before been incorporated into a regulated social system. Alternatively they have become open access resources through institutional failure that has undermined former collective or individual management regimes. For example, numerous collective management regimes to natural resources were undermined through colonialism, leaving the resources without defined access. In the case of an open access resource, individuals have both a privilege and no right with respect to rates of use and maintenance of the resource. Cetaceans in the $19^{th}$ century were regarded as open access resources, as they belonged to no one until they were captured (Bromley 1991; McCay 1996; Hanna 1999).

## Institutions

Common property regimes, as collective resource management systems, have been shown to develop when a group of individuals is highly dependent on a resource and when the availability of the resource is uncertain, limited or requires large-scale cooperative community effort to harvest (Runge 1992). If the resource problem is repeatedly experienced, such as a low or zero catch, and if it exists within a single community of users, the users are likely to develop a collective institutional arrangement to deal with the problem. The reliance on wide community cooperation to ensure a successful catch also ensures the users develop a collective institutional arrangement to ensure community participation and an equitable share of the catch.

The term 'institution' is used to describe the bodies of rules, regulations and processes that guide the decision-making of resource users. Such decision-making is often carried out within organisational structures made up of groups of people acting in relationships governed and legitimised by institutions (Ostrom 1990; Ostrom *et. al.* 1992; Alcorn and Toldeo 1998).

Institutions are defined as humanly-devised constraints that structure human interaction. They are made up of formal constraints (rules, laws, and constitutions), informal constraints (norms of behaviour, conventions and self-imposed codes of conduct) and their enforcement characteristics (North 1993). They are also described as locally devised systems of property rights and obligations that are culturally anchored in a detailed time and place. Institutions define the boundaries of the resource, its users, the rules and regulations, and processes for use of the resource. Institutions form complex social phenomena, not only among people but also between people and the resource. Within institutions we often see complex social relations played out which protect the cultural norms and self-interest of the resource users and their resource (Ostrom and Schlager 1996). Thus:

> Institutions result from a political process of trade-offs between members of a community who must work together because of their interdependencies in many other spheres. The institutions may be old, but they are dynamic and respond to change

according to interpretations of local law as it is applied in specific circumstances (Alcorn and Toledo 1998:220).

**Institutional Performance**

Institutions have to deal with two fundamental management problems that arise from the basic characteristics of all resources: how to control access to the resource (the exclusion problem), and how to institute rules among users to solve the potential divergence between individual and collective rationality (the subtractability problem) (Berkes and Fölke 1998). Because of the limited number of property rights regimes, it is a relatively easy task to determine what type is in operation for a particular resource. However, when describing institutions, the task is far more complex, because analysis must include the historical, political, cultural, social and environmental attributes of each institution under consideration. While these exercises are important in understanding how people use resources, such an exercise does not inform us of institutional performance or provide evidence of sustainability of the resource and the institution itself.

In their discussion of indicators for sustainable common property institutions, Gibbs and Bromley outline the following criteria from which they assert institutional performance can be evaluated:

- a minimum (or absence) of disputes and limited effort necessary to maintain compliance: the regime will be efficient.
- capacity to cope with progressive changes through adaptation, such as the arrival of new production techniques: the regime will be stable.
- a capacity to accommodate surprise or sudden shocks: the regime will be resilient.
- a shared perception of fairness among members with respect to inputs and outputs: the regime will be equitable (Gibbs and Bromley 1989:26).

Using the criteria of efficiency, stability, resilience and equability, Gibbs and Bromley assert: "the institutional performance of such regimes can be appraised and their value as foundations for institutional innovation can be assessed" (*op. cit*). Moreover, "[i]n order to meet these criteria and to guide the use of collective resources, numerous variables must be defined by the group implementing common-property resource management" (*op. cit*). These variables include:

- what constitutes membership of the group having a right to the resources and who is not a member with the duty to respect the right of the members?
- what constitutes agreement—unanimity, consensus or majority?
- on what basis the right will apply over time, *i.e.,* annually, seasonally?
- how rights are transmitted between generations.

- where control resides, *i.e.,* vested in a community board, in village or district elders, or in households.
- how compliance with agreed rules and conventions is maintained.
- how departures from the rules are to be corrected and sanctions imposed.
- how disputes are resolved (*ibid*: 26-27).

Researchers involved in the task of identifying whether or not these criteria are evident in groups implementing common-property resource management have identified essential elements or conditions that help account for success of the institution and the maintenance of the resource itself; these essential sets of elements are called 'design principles.'

**Design Principles**

A design principle is one of a number of essential elements or conditions that help account for the success of these institutions in sustaining the common property resources and gaining compliance of generation after generation of appropriators to the rules in use (Ostrom 1990:68-69), and at this time:

> we are beginning to understand the design principles of robust institutions. Robust institutions are found in systems that have survived for long periods of time. Such institutions survive when operational rules are adapted in relation to a set of collective choice and constitutional choice rules... . Fragile institutions, as suggested from evidence from the field, tend to be characterised by only some of these design principles and failed institutions are characterised by only a few (if any) of these principles (Ostrom and Walker 1997:46-47).

When fragile and failed institutions can no longer manage resources sustainably, they threaten both the resource and the resource-using community. However, uncertainty remains regarding the relationship existing between particular design principles and institutional success:

> I am not willing to argue that these design principles are necessary conditions for achieving institutional robustness in common property resource settings... . I am willing to speculate, however, that after further scholarly work is completed, it will be possible to identify a set of necessary design principles and that such a set will contain the core of what has been identified here (Ostrom 1991:90-91).

For the design principles to constitute a credible explanation for the persistence of common property resources and their related institutions, Ostrom asserts that one will "need to show that they can affect incentives in such a way that appropriators will be willing to commit themselves to conform to operational rules devised in such systems, to monitor each other's conformance, and to replicate the common property resource institutions across generational boundaries" (*ibid*: 91). Design principles associated with long-enduring common property resource institutions are identified and their operational features described in the table below:

**Table 1. Design principles associated with long-enduring common property resource institutions (Ostrom 1990:90)**

| | |
|---|---|
| Clearly defined boundaries | Individuals or households with rights to withdraw resource units from common property resource and the common property resource itself are clearly defined. |
| Congruence between appropriation and provision rules and local conditions | Appropriation rules restricting time, place, technology, and/or quantity of resource units are related to local conditions and to provision rules requiring labour, materials, and/or money. |
| Collective choice arrangements | Most individuals affected by operational rules can participate in modifying operational rules. |
| Monitoring | Monitors, who actively audit common property resource conditions and appropriate behaviour, are accountable to the appropriators and/or are the appropriators themselves. |
| Graduated sanctions | Appropriators who violate operational rules are likely to receive graduated sanctions (depending on the seriousness and context of the offence) from other appropriators, from officials accountable to these appropriators or from both. |
| Conflict resolution mechanisms | Appropriators and their officials have rapid access to low-cost, local arenas to resolve conflict among appropriators and officials. |
| Minimal recognition of rights to organise | The rights of appropriators to devise their own institutions are not challenged by external government authorities. |

And for common property resources that are part of larger systems:

| | |
|---|---|
| Nested enterprises | Appropriation, provision, monitoring, enforcement, conflict resolution, and governance activities are organised in multiple layers of nested enterprises. |

***Clearly Defined Boundaries***

In any efficient system for managing resources, clearly defined boundaries, along with specifying the authorised users to operate within the boundaries, are often thought of as the first steps. "So long as the boundaries of the resource and/or the specification of individuals who can use the resource remain uncertain, no one knows what is being managed and for whom" (*ibid*: 91). A resource without defined boundaries and identified users is an open access resource and with the absence of property rights and no ownership assigned, that situation can lead to 'free riding' and resource over-exploitation. Not only are there sustainability reasons, but also sound economic reasons for clearly defining boundaries: "Without defining the boundaries of the CPR [common property resource] and closing it to 'outsiders,' local appropriators face the risk that any benefits they produce by their efforts will be reaped by others who have not contributed to those efforts" (*op. cit*). For users of the resource to maintain interest in managing the resource and contributing their time, energy and resources, defined boundaries and identified users are essential to the longevity of the resource and, in some instances, of the community of users:

> If the aim of the resource users is sustainability and this is the case with all users of flow resources, then closing the boundaries is not sufficient to guarantee such an outcome. It is still possible for a limited number of appropriators to increase the quantity of resource units they harvest so that they either dissipate all potential rents or totally destroy the resource. For this reason, it is therefore necessary to implement additional rules of use and management (*ibid*: 92).

***Congruence Between Appropriation and Provision Rules and Local Conditions***

The second design principle is 'a match between contribution and benefits' (Weinstein 2000:381):

> Community-based management requires substantial investment from all participants. This comes as a mixture of resources and time. All long-enduring regimes require considerable amounts of time and effort by the membership. To be long-enduring, people need to 'stay in.' That is, they need to remain actively involved, and continue investing in the process. Consequently, participants must feel that they are necessary [and] for the members to feel that the process is imbued with equity and fairness (*op. cit*).

Appropriation problems are those concerned with the allocation of the flow of the resource and provision problems associated with the resource stock. These two problems are always linked to some degree in common property resources, and depend on the particular configuration of variables related to the physical world, the rules used, and the attributes of the individuals involved in each specific setting:

> When appropriators face appropriation problems, they are concerned with the effects that various methods of allocating a fixed, or time-independent, quantity of resource units will have on the net return obtained by appropriators... . Provision problems concern the effects of various ways of assigning responsibility for building, restoring, or maintaining the resource system over time, as well as the well-being of the appropriators (Ostrom 1990:47).

From the common property literature, it is evident that locally-devised use and process rules aid in the sustainability of common property resources (*e.g.,* Davis 1984; Berkes 1989; Young *et al.* 1994; Berkes and Fölke 1998).

***Collective Choice Arrangements***

This design principle concerns the right to make rules or change rules being firmly held by the resource users or their own community's local institution. Consequently, any ability to make changes is also limited to those affected by the changes. Weinstein (2000:381) points out that this is a solution to one of the problems of the top-down style of management, where resource users who feel that changes are not made in response to local conditions, or that changes do not match their needs, may feel justified in ignoring the rules.

Again, there are strong economic incentives contained in collective choice arrangements. Creating a new collective choice arrangement can sometimes be easy and involve little cost, particularly when local participants can modify operational rules. However, such positive outcomes are less likely when external managers or advisers, who know little of local history, culture and circumstances, are brought in to determine and/or impose new rules to solve problems. Collective choice arrangements also provide a forum where local knowledge is understood and valued by rule makers. There is also a sense of ownership of the rules that are developed about resources and often greater likelihood that the rules will be complied with, and that the community will not have the added burden of costly enforcement procedures.

Common property resource institutions with this design principle are better able to tailor their rules to their own specific cultural norms, environmental conditions and other local circumstances as determined by their own members. However, it cannot be claimed that just because the users or the community itself developed the rules that they will be adhered to by all users, for as noted: "Nor is the fact that appropriators themselves designed and initially agreed to the operational rules in our case studies adequate explanation for centuries of compliance by individuals who were not involved in the initial agreement" (Ostrom 1990:93). What is evident from the case studies discussed by Ostrom (*ibid*) is that even in repeated settings where reputation is important and where individuals share the norm of keeping agreements, reputation and shared norms are insufficient by themselves to produce stable cooperative behaviour over the long term. Each of the case studies found it was evident that to aid collective choice arrangements, active investment in monitoring and sanctioning activities is also required.

***Monitoring***

This design principle is self-explanatory and, as with other design principles, there may be clear economic benefits (and incentives) to the user community when monitors are themselves resources appropriators or otherwise members of that community, allowing monitoring to be undertaken in the course of resource use with no extra equipment or personnel needed to undertake the task. Top-down policy makers, however, are often cautious of implementing this type of monitoring system as they see it as a case of 'the fox looking after the hen house,' and therefore open to potential abuse. However, as Weinstein

(2000:382) observes: "In addition to oversight, the monitoring may also include biological and physical information important to management decisions. In this case, monitoring is used to collect information useful to the community of users. The resource users make decisions about precisely what information would be useful." Instances such as this, which provide user incentives, assist with gaining wider community support for the common property rights framework.

***Graduated Sanctions***

Some degree of violation of the rules occurs in any management system and has the ability to burden the resource users with additional costs or, if serious, threaten existence of the management institution and/or the sustainability of the resource. Like the other design principles in the case studies, the resource users themselves, or their community, undertake this responsibility. The first infraction may be met with a relatively low sanction, such as a warning not to repeat the offence, while continued infractions may be met with harsher sanctions that may include confiscating equipment for a period of time. In the most extreme cases, infractions may see the offender expelled from the resource-using group and their rights terminated (*op. cit*).

***Conflict Resolution Mechanisms***

Conflicts are an inevitable characteristic of any governance system and are often due to ambiguity of rules or varied interpretations of the rules (*op. cit*). If individuals are going to follow rules, either over the short or long term, there needs to be some mechanism for discussing and resolving what constitutes an infraction of the rules so that any small dispute does not escalate and threaten both the resource and the appropriators. "In theoretical models of rule-governed behaviour, the rules that structure the strategies available to participants are unambiguous and are enforced by external, all-knowing officials. In field settings, applying rules is never unambiguous, even when appropriators themselves are the monitors and sanctioners" (Ostrom 1990:100). In practice, people often find different ways to interpret the same rule to benefit themselves, thus to avoid disputes, conflict resolution mechanisms are highly desirable.

Gibbs and Bromley's (1989:26) criteria for measuring institutional performance included a minimum (or absence) of disputes and limited effort needed to maintain compliance and assist the regime to remain effective, and as such, it is an important indicator of whether or not a common property resource institution can be considered sustainable.

***Minimal Recognition of Rights to Organise***

Resource users frequently devise their own rules without creating formal governmental jurisdictions. The rules and processes of common property resource use typically develop slowly, sometimes over centuries, growing up around day-to-day practice, influenced by history, culture and biological or environmental conditions, until reaching a stage when the day-to-day use could be described as a set of formal or informal well-developed rules. Such rules can determine who gets to use a resource, in what season resources are used, the number and size harvested, equipment used and how the resource may be divided between the users and their communities. When small local communities of indigenous or local peoples are in a situation where they find themselves governed by outsiders, well-established rules used for long periods of time may become threatened when external governments and their officials fail to recognise what works best in practice.

What we see from the case studies is that, provided external government officials give at least minimal recognition to the legitimacy of such customary rules, the resource users themselves may be able to enforce the rules. However, when government officials choose to ignore long-established rules, the incentive for the local resource users is also weakened or removed altogether. And, as Ensminger (1996) demonstrated in her African case study, when no recognition is given to traditional resource-use systems, the result is often the breakdown of the system itself, which often results in a secondary failure when external systems are imposed upon that society in an effort to correct the initial breakdown.

***Nested Enterprises***

Because "locally devised systems of property rights and rules are anchored in detailed time and place, cultural norms and self-interest of resource users" no two case studies will ever be identical (Ostrom and Schlager 1996:146). To outsiders, these complex property rights regimes and institutions are often considered too complex, expensive or a relic of the past that cannot meet contemporary pressures of the global economy.

An alternative approach that holds the promise of supporting local-level institutions, retaining the benefits of such institutions, and addressing their limitations, is to nest these organisations in a larger institutional framework that facilitates co-ordination among them, and can address large-scale problems that local-level organisations share in common:

> As problems become more persistent and complex, involving larger numbers of local organisations, a more formal, umbrella organisation may be created to better co-ordinate the actions of its members and to address shared problems that no individual member can address alone. To protect the integrity of local-level organisations, such nesting is best accomplished through a federal and not a hierarchical approach. Each local-level organisation retains exclusive jurisdiction and serves as a mechanism to support and enhance the viability of local-level organisations, and to exercise authority in areas that its member organisations have ceded to it. Just as individuals collaborate to devise local-level, self-governing organisations, so too do local-level, self-governing organisations devise regional-level organisations (*ibid*: 148).

The manner in which various design principles (*see* Table 1) help ensure institutional performance has been commented on as follows:

> When CPR appropriators design their own operational rules (design principle 3) to be enforced by individuals who are local appropriators or are accountable to them (design principle 4),

using graduated sanctions (design principle 5) that define who has rights to withdraw units from the CPR (design principle 1) and that effectively restrict appropriation activities, given local conditions (design principle 2), the commitment and monitoring problem are solved in an interrelated manner (Ostrom and Schlager 1996:99).

# *Historical Overview of the Political Economy of the Faroes*

> *Anyone can take part in a whale drive and get their own share of the food, but even those who don't take part also get a share of the food. Everyone in the communities gets something. For me whaling is a part of who we are—not only because it's what we eat—but also because it's something that keeps our communities close. This means a lot, in this modern computer age, and we are proud that we have been able to keep this way of life.* Jørgen Niclasen, Faroese Minister of Fisheries (WCW 1999:10).

The defining characteristic of world-system analysis is that the appropriate 'unit of analysis' for the study of social and or societal behaviour is a 'world-system' which highlights dynamics between local, national and global spheres and denies that the 'nation state' can be viewed as a relatively autonomous 'society' that 'develops' over time (*op. cit*). In this regard, political economy and world-systems analysis is used to examine the 'macroview' of Faroese society and history from the time of Norse settlement through to the present day. Within this macro-perspective, I will examine how the development and adaptation of the *grindadráp* through time has been shaped by the political economy of the Faroes and how the Faroese have responded to these external forces in an effort to secure greater decision-making over their food resources. Thus an examination of long-term and large-scale social change will inform the 'microview' of the *grindadráp* itself, and will be examined later in the book.

Although this historical overview identifies four periods of time, it is important to remember Wallerstein's observation that major large-scale social processes, such as incorporation into the capitalist world economy, are not abrupt phenomena, but rather:

> They emerge from the flow of ongoing continuous activities. While we may give them dates retrospectively and approximately, the turning points are seldom sharp and the qualitative changes they incarnate are complex and composite. Nevertheless, they are real in their impact and eventually they are perceived to have occurred (Wallerstein 1989:32).

The first period extends from the beginning of Norse settlement of the Faroes in the early 9th century through to the mid-16$^{th}$ century, during which time Faroese political relations were initially closely tied to Norway and then Denmark. The second period extends from the 16$^{th}$ century through to the early 19$^{th}$ century, with the emergence of the Faroese commercial fishing industry. The third runs from the early 19$^{th}$ century through to the beginning of the 21$^{st}$ century, during which time Faroese economic reliance on marine products reached almost 96 percent of exports (*Hagstova Føroya* 2000:8). The fourth and last period covers the early years of the 21st century.

Within each of these historical periods, the emergence of property rights to whales and the local institution that slowly developed around the practice of whaling to protect those property rights, is considered. This historical overview of the political economy of the Faroes speaks to the question of the inappropriate portrayal of the Faroes as a discrete cultural, economic and political isolate wherein unchanging and outdated practices continue to operate from earlier times.

## The First Period: 800 AD–1500

Little is known of the early period of settlement of the Faroes, and what is known has been derived from early written accounts and the archaeological record.

Evidence from an archaeological excavation of a Viking Age homestead at Kvívík, on Streymoy, indicates that the early settlers were utilising pilot whales and other local marine resources as an important source of food:

> Part of the diet of those who lived at Kvívík is revealed by finds of bones of sheep, cows, pigs, seals and pilot whales, guillemots, razor-bills, cormorants, sea-gulls, and, of course, cod. It is uncertain whether they ate horses, but bones tell us they certainly had them, and we know that they gave toy horses to their children (Dahl 1971:69).

It is not clear how the early settlers obtained pilot whales, perhaps from either strandings, which continue to occur today, or from an early form of drive fishery. There are different views concerning the origins of Faroese pilot whaling. Some suggest the pilot whale drive fishery may not have originated in the Faroes, for it is reported that Iron Age settlers arriving in the Orkneys from the Hebrides "were particularly skilled in whale hunting—driving schools of whales into bays" (Hamilton 1964:75). This report suggests specific whaling methods were already established in the Orkneys at the time of Norse settlement there, and may have come to the Faroes with the Norse settlers arriving from the Orkneys and Shetland. Others suggest that the driving of small whales and the harpooning of individual whales at sea were established practices along the coast of Norway at the time of the Norse expansion, and may have come directly to the Faroes from Norway (Johnsen 1962: translated in Sanderson 1992:27). Recent research, however, proposes two reasons why Faroese pilot whaling is most likely a local innovation. First, pilot whales were relatively rare in Norwegian coastal waters, and catches were too infrequent to find an institutionalised form. And, second, while pilot whales in the Faroes were driven ashore, they were in most places in Norway enclosed by seines. The difference had important socio-economic implications. Technology as well as regulations and practices pertaining to pilot whaling and distribution of the catch in the Faroes must therefore be understood in its local context rather than as an established institution brought by settlers from Norway (Kalland 2010:100).

Other archaeological finds included spindle whorls and loom weights, suggesting spinning as a major occupation and important both domestically and for international trade. The Faroese, at this time, were known to trade with Norway and Britain, exporting woollen garments for necessities such as grain, soapstone, timber, glass and amber beads, and hazel nuts (Wylie 1987:9). The Faroese Sagas (*Trøndur of Gøtu, see* Johnston 1994) suggests some of the more important farmers owned ships, which enabled them to trade with Norway.

The Faroese economy at this time was agricultural, based on landholders raising sheep and cattle and some barley—although due to the climate, barley was harvested green and dried over peat fires—and subsistence pursuits such as wildfowling, whaling, and fishing provided important food resources should the crops fail in the harsh North Atlantic climate.

The discovery of pilot whale bones and teeth, along with the skeletal remains of sheep, cattle, fish, and seals, unearthed at another archaeological site in Leirvík on Eysturoy dating from the $12^{th}$ to the $13^{th}$ centuries, demonstrate that whales were an integral part of the Faroese economy at that early period of history (Arge 1997). Considering the poor soils and climatic constraints, inhabitants of the Faroes have always been dependent on marine resources together with imported grain, timber and other commodities.

At this early period of settlement, the Faroes were predominantly a self-governing society. A study of the Faroese Sagas (*Trøndur of Gøtu*), concludes that soon after the Norse settlers in the Faroes organised themselves into a community they then sought to remain independent of Norwegian suzerainty (Hansen 2003:101). This, Hansen (*op. cit*) proposes, is the first stage of the settlers creating a distinctive cultural identity in which they came to regard themselves as Faroese. The seat of government was situated then, as it is now, on the rocky spit of Tinganes situated on the harbour of Tórshavn. It is here that the annual legal assembly, the *Løgting*, would meet in late spring or early summer. According to Wylie: "The central parliament may at one time have been a popular assembly, to which all free men might come; but men probably attended less on their own behalf than as followers of regionally powerful chieftains" (Wylie 1987:9).

Norway began to extend influence over the Faroes, regularly collecting taxes, around the year 1035. It was at this time that the Faroese became Christian, acquiring their own diocese with an Episcopal residence established at Kirkjubøur. Both the church and the Norwegian Crown had representatives in the *Løgting*, which although the highest court in the Faroes, gradually had its powers reduced to interpreting laws dictated by the Norwegian Crown (*op. cit).*

Sometime before the year 1200, the Norwegian King, in an attempt to regularise the administration of his scattered realms, began appointing an important man in each district to represent him and his law. The King's chief representative on the Faroes was called the bailiff (*fúti*), and under him there were six sheriffs (*sýslumenn*), one for each of the six districts (*sýslur*) in the Faroes. While the bailiff was a foreigner, the sheriffs were Faroese and generally lived in the district they administered (Wylie 1987:11).

Slowly, as the Norwegian King asserted more control over the Faroes, a system developed where the bailiff would arrive in the Faroes on the first spring trading vessel and leave again for Europe at the onset of winter, leaving in place a local representative who was known as the 'winter bailiff' (*vetrarfúti*). The bailiff's core responsibilities included the collection of taxes and prosecuting cases before the *Løgting*. The bailiff was the principal connection between the Faroes and Europe, and over time his powers increased to include appointing representatives (*løgrættumenn*) to the spring parliament, the *Løgting*. Early Faroese law was recognised by the Norwegian King (Magnus), who in 1273 wrote:

> We want all men to know, that we have agreed to confirm for you, according to your request and according to the advice of all the best men, that such laws shall be valid here as are valid in the whole of the *Gulating* district [south-western Norway]. Except that chapter of Laws concerning land shall stand according to what your own book already witness (Jakobson 1907:23-24, cited in *op. cit*: 11-12).

In 1277 Faroese decision-making was further reduced when the Norwegian King imposed his law on the Faroes, taking the law-making power out of the hands of the Faroese *Løgting*. Also around this same time, the King assumed the right of confirming the election of the foremen of the *Løgting*, the *løgmaður* (*ibid*: 12).

The control of trade was also slowly removed from the Faroese and carried out by foreign interests, largely the King of Norway, who attempted to regularise and share in the profits from the North Atlantic trade (*ibid*: 17). In the late $13^{th}$ century the Norwegian king banned Hanseatic[1] merchants from carrying out business in the Faroes, securing his control over the islands. In return, the King agreed to send two vessels per year to the islands for trade purposes, thereby establishing a *de facto* trade monopoly, keeping the Faroese dependent on the rulers of Norway (*ibid*: 18).

Norway's hold on the Viking settlements began to decline in the $13^{th}$ century, with the Hebrides and the Isle of Man passing to Scotland, and Norway's focus of trade shifting to the south and east, especially to the ports of Northern Germany. Relations with settlements in Northern Britain grew weaker. In 1299 the Norwegian capital and royal chancellery were moved from Bergen to Oslo. A century later, in 1397, under the Treaty of Kalmar, Norway was formally united with Sweden and Denmark. While the union with Sweden was short, the union with Denmark lasted until 1814 and the Faroese "became a more peripheral part of the realm, increasingly different from their rulers and the foreigners with whom they came in contact by way of trade" (*ibid*: 18).

Norway, now governed by Denmark, ensured its former island dependencies in the North Atlantic grew less important. In 1468 and 1472 respectively, Orkney and Shetland became part of Scotland, with the Faroes and

---

[1] The Hanseatic League was an alliance of trading guilds that held a trade monopoly over the Baltic Sea and to parts of the North Sea and the majority of Northern Europe from the $13^{th}$ to the $17^{th}$ centuries (*see* Wylie 1987:17-18)

Iceland remaining a part of the Danish realm while becoming more politically isolated from Europe.

**The Emergence of Property Rights over Whales**

For a discussion of the emergence of property rights to whales during the early period of settlement recorded in the earliest written references to whaling in the North Atlantic, I rely on Sanderson (1992) whose comprehensive work translates old Norse texts and more recent texts in Norwegian and Danish. Her work is an examination of "what representations of whales and whaling in the Faroes reveals about attitudes to the natural world, and in particular, to the *grindadráp*—the driving and slaughtering of schools of pilot whales for food—and its significance in Faroese culture and economy from the earliest time until the 19th century" (Sanderson 1992:2).

As noted above, the opportunistic sighting, driving and slaughtering of pilot whales is not the exclusive domain of the Faroese, rather it is rooted in the traditions of several Nordic North Atlantic communities. For many hundreds of years, pilot whales were hunted in the North Atlantic—in Norway, Iceland, the Orkneys and Shetland—and like the Faroes today, pilot whales were driven into bays and slaughtered (Williamson 1948).

With the Faroese conversion to Christianity at the beginning of the 11th century, whale meat became doubly valuable as the church permitted the eating of whale meat during Lent and other periods of fasting, while forbidding consumption of the meat of land animals and marine mammals such as seals (Sanderson 1992).

Old Norse legal codes of 1281 recognised the importance of whales to the economies of medieval Iceland, Norway and the Faroes. The legal codes, known as *grágás* and later *Jónsbók,* outlined rights to and the division of driftage, especially washed up or stranded whales. The general rule, of the *grágás*, which remains in Iceland today, was that:

> Every man owns whatever drifts up on his land, of wood and seals, and whales and fish, unless the rights of the foreshore have been sold from the adjoining land or given away or handed over in payment and then whoever has the authority thereafter has the rights to the foreshore (*Grágás* 1852, Vol I:123; translated in *ibid*: 31).

The law recognised the labour of the hunter, even if he lost his catch and it later washed ashore. The finder was obligated to notify the owner within a day if the carcass still held a marked harpoon and he therefore knew to whom it belonged. Whales were often harpooned close to shore and if they could not be taken they were left adrift in the hope that they would wash ashore where they could be claimed later (*ibid*: 32). The *grágás* stipulated: "If men drive whales ashore they own two-thirds (of the catch) and the landowner one third" (*op. cit*).

The earliest known reference to whales and whaling in the Faroes was made in 1298 and is contained in the *Seyðabrævið*, or what is known as "The Sheep Letter." The *Seyðabrævið* was a special enactment regulating the use of land, especially the joint use of outfield pastures, exclusively concerning the

Faroes. Duke Hákon Magnússon issued this edict as an addition to the *Gulathing* Law, which operated in the western provinces of Norway, and was made effective during the reign of Magnússon's father, King Magnús the Law Amender, in 1273. The *Gulathing* Law outlines in detail the shares of both stranded whales and whales which were hunted to which hunters and landowners were entitled (*ibid*: 27).

The *Seyðabrævið* contains two brief paragraphs dealing with rights to shares of stranded whales, whales driven ashore, and whales found at sea. Sanderson (*ibid*: 27-28) suggests these laws must be seen as changes or exceptions to detailed and complicated regulations for rights to whales outlined in the *Gulathing* Law, which also applied to the Faroes:

> The first paragraph [of the *Seyðabrævið*] stipulates the share allotted to farmer's servants who find a whale at sea and cut off a boat-load, in which case their share is one seventh. It also states that if they bring a whole whale ashore uncut, their share is that that is outlined in the *Gulathing* law (i.e., one half). If men drive a whale ashore, kill it and secure it above the high water mark on land that is not theirs, they are entitled to one quarter (the same as driftwood, sheep or cattle or any other animal found at sea and brought ashore). A further paragraph refers to 'killer cuts' (*vagn hogg)*, i.e., pieces of whale which are found as driftage, and these then belong to the owner of the land where they drift ashore (Poulsen and Zachariassen 1971:49-51, translated in *ibid*: 28).

It was apparently necessary to stipulate the share for a farmer's servants as one- seventh of anything cut off a whale found at sea, according to Bjørk because "if the farmer, who would also be the owner of the boat, received the entire share, there would be little incentive for his servants to procure such a valuable source of food, nor, it may be added, to exert the collective effort necessary to secure it" (Bjørk 1963:III,171, translated in *op. cit*).

Sanderson notes that the brief extracts from the *Seyðabrævið* provide no information about the degree of organisation that existed at this time for driving whales ashore. Some commentators consider this reference in the *Seyðabrævið* refers to the *grindadráp* (the pilot whale drive), however Sanderson asserts this is not explicit in the text and she is of the opinion that it more likely refers to:

> the catching of individual whales of larger species such as right or fin whales, than driving ashore entire schools of small whales. The reference to the whale is in the singular, and one of the stipulations is that the whale be secured above the high water mark, a procedure which is relatively simple with small three or four metre toothed whales, but quite a different undertaking with, for example, a 15 or 20 metre baleen whale (*op. cit*).

To be successful, a *grindadráp* must involve a significant number of people capable of being quickly and effectively highly organised. With this in mind, Sanderson (*ibid*: 29) suggests it is difficult to conclude to what extent it may have been possible to organise a whale drive, particularly with the intricate system of message sending by means of runners, the use of smoke signals, and the participation of the many boats needed in order to be effective. While the extent of boat ownership at this time was unknown, what is known is the population of the Faroes at the beginning of the second millennium was very small, estimated to be somewhere in the vicinity of 3,000 to 4,000 spread across the islands in small villages of up to 100 people (Wylie 1987:14). If the exact method of taking whales in the Faroes in the 10th, 11th and 12th centuries remains unresolved, Sanderson says, "it can be said with some degree of certainty that the pilot whale, as well as other common occurring whale species, have constituted a significant part of life and the local economy in the Faroes since the time of Norse settlement of the islands in the 10th century" (Sanderson 1992:29).

The earliest illustration of whaling in the Faroes appeared in 1539 in Olaus Magnus' atlas *Carta Marina* (Map of the Northern World) wherein an illustration depicts the Faroes with a whale hooked and securely fastened in the process of being flensed (Fig. 6). In an explanation accompanying the illustration, Magnus refers to the Faroes as a place "where the inhabitants eat fish and cut up the great whales which are stranded in bad weather, and divide them amongst themselves" (Nørlund 1944, translated in *ibid*: 46).

**Figure 6**. *A 1539 map of the Faroes by Olaus Magnus* (from Sanderson 1992).

What we see in the *Seyðabrævið* and other early Norse legal codes operating in Iceland, Norway and the Faroes, is the emergence of formalised property rights. Whales, as we see from Sanderson's discussion of the early Norse legal codes, were so highly prized that it was necessary to clarify in great detail, through a binding legal instrument, what rights and shares the 'owners' of the whales would receive under what circumstances in advance of finding the whale(s). These early property rights have several functions. They outline the population of legitimate owners, *viz.,* the landowner, the boat owner(s), and those involved in securing the whale (the servants). They also specify the allowable actions of these various owners and their associated responsibilities so that expectations are consistent and enforcement of the rules is possible; consequently, before the whale is secured, each party knows what their share will be and this ensures their collective effort is rewarded. As whaling or the retrieval of drift whales required the labour of many people, some incentive had to be provided to ensure participation and cooperation. The early literature on Faroese pilot whaling provides evidence of the emergence of design principles (to be further discussed below), including the match between contribution and benefit of those involved in whale procurement and distribution, and the emergence of defined boundaries identifying individuals possessing rights to withdraw resource units from the common property resource—all shaped by the wider political economy and the class systems these rules served to maintain. The recognition of property rights significantly reduces the costs of conflict and uncertainty, and in a wider sense connects elements of the natural system (the whales), to elements of the human system (the local communities). An understanding of the social history of the early second millennium clearly indicates these property rights are rooted in the social and cultural norms of the day.

## The Second Period: 1540–1800

The early 16th century saw Denmark in chaos following the outbreak of civil war. The fractious relationship between the nobles and monarchy aided the success of the Reformation in Denmark, which in turn ensured a transition in the Faroes and the reconfiguration of land tenure:

> In 1537 Christian III abolished the office of the archbishop and took over the governance of the church. Piously and in order to fill the depleted royal coffers, he appropriated the third of the tithe that had previously gone to the bishops and the lands belonging to the church, at least a third of all the cultivable land in the realm (Wylie 1987:20).

When the Danish Crown took over the Church's land its holdings increased about tenfold. This land was then let to tenant farmers, who were known as 'King's farmers' (*kongsbøndur*). During this period, the bailiff's powers also increased; as well as collecting taxes, he was also charged with renewing leases. This led to corruption, with the Faroese claiming the bailiff was evicting them from their farms without 'good reason.' Laws were changed to prevent such corruption, but over a century later the bailiff was selling tenancies when they became vacant to the highest bidder. The law was again changed to prevent continued corruption, with Crown tenancies now being passed from father to son. While the *Seyðabrævið* required leases to be held from year to year, it seems they were held for much longer periods, perhaps for a tenant's lifetime (*ibid*: 32-33).

By the 17th century, a system had grown whereby the King of Denmark controlled all the trade by renting out trading privileges to merchants, sometimes also with exclusive rights to collect taxes. The system held the potential to become truly repressive, with the Faroese held to ransom at the whim of continental merchants (*ibid*: 21-22). Unlike the situation in Denmark, the Faroes had no nobility or middle class, and the bureaucracy was almost entirely held by the office of the bailiff:

> The inherently small capacities of land and the crown's increasing restrictions on the marginally profitable Faroese trade isolated the islands further, forestalling economic growth and the rise of the native middle class. The Faroese remained free peasants, whose way of life changed little until the 19th century (Wylie 1987:22).

In an economic sense, the way of life for the Faroese may have changed little from early settlement until the 19th century. However, the colonial processes and their subsequent social changes further challenged Faroese identity. During the 17th century, Danish became the language of the church, state, and trade, with the *Seyðabrævið* being translated into Danish in 1637. Faroese was becoming a language of the house, neither used by the state nor the church.

The subsistence mode of production and the small amount of trade goods it produced contributed to the maintenance of the Faroese language, which survived through small Faroese institutions shaped by economic life, such as *kvøldsetur*—the telling and listening to folktales and legends often in the kitchen while families and hired hands would card, spin and knit wool or do other chores—which has been referred to as 'a primary institution of remembrance' (*ibid*: 41), and remained an established part of Faroese family life until the late 19th century, when economic change rendered such practices obsolete. The Faroese language was also kept alive through singing and dancing to heroic ballads (*kvæðir*)[2] and satiric ballads (*tættir*)[3] on special occasions, including at the end of a *grindadráp* where the men would dance themselves warm while fresh whale meat was prepared and the sheriff and his helpers carried out the assessment and the division of the whales among the people. Others have written on the place of traditional songs in creating a Faroese national identity (*e.g.*, Wylie and Margolin 1981:129; Joensen 1990:182; Sanderson 1992:86; Kalland *et al.* 2005:125).

---

[2] *K* deal mostly with the distant past and can be over 100 verses long. "They reach back through the Viking Age and the Faroes conversion to Christianity (*Sigmundarkvæði*) to the Charlemagne cycle (*Rólantskvæði*) and Germanic myth (*Sjúrðarkvæði)*, about Sigurd (the Dragon-Slayer). Most are based on written sources and seem to have been composed in the late Middle Ages (Wylie 1987:43).

[3] *Tættir* deal mostly with local figures in a satirical vein, and are sung and danced to at festival times in the same way as *kvæðir* (*op. cit*).

Wylie (1987:26) describes the Faroes at this time as having two economies. The first is a highly diversified domestic economy, where the men fowled, whaled and fished, cut turf, grew hay and some barley, and kept sheep and cattle. This domestic economy was supplemented by the external economy, where merchants bought tallow, feathers, sheepskins, butter, whale oil, and dried fish. The most important item of trade was wool, particularly when woven into stockings, which received a better price than raw wool. The domestic economy therefore is identical to what Wolfe describes as a subsistence society, one that includes, in addition to commercial trade, a domestic mode of production involving a seasonal round of production, networks for sharing, traditional systems of land/water use and occupancy, and inter-generational systems of belief, knowledge and values associated with resource uses (Wolfe 1983:272).

The economic order that the Faroese endured during this period did not arise painlessly by itself. It was imposed and maintained by force, and for this reason it is described as a political economy. The subsistence economy was important because the trade system became repressive with the tightening of Danish control over Faroese trade. Moreover, this control was not just over trade, but also over communication with the outside world, local government, justice and the collection of rent and taxes. The Danish trade monopoly "was a bleak, bitter and often hungry time... its maintenance entailed the Faroes' continued economic stagnation and the survival until well into the nineteenth century of a society essentially unchanged since about 1620" (Wylie and Margolin 1981:5). The internal diversified subsistence economy based on the common property resources of whales, fish, and fowls kept the Faroese communities alive by providing a relatively reliable source of local low-cost food.

The abolition of the trade monopoly was officially proposed in 1789, with some opposition from a small number of Faroese farmers whose own privileges were to be diminished. However, because of the increasingly unsettled state of European politics at the time, it was not carried out by the Danish Crown. This saw a revision of the price schedule, which encouraged the continued production of stockings and "discouraged what eventually proved to be the key to the Faroes' economic development, the commercial exploitation of the sea" (Wylie 1987:67).

During the 1770s some Faroese were trained in the manufacture of *klipfish* (dry salted cod), but this period of time witnessed poor fishing seasons and the manufacture of *klipfish* did not succeed. The Faroese economy remained dependent on its small agricultural base as evident in the statistics of Faroese exports between 1767 and 1776, with 96.9 percent of exported goods being wool, woollens, tallow and lambskins (Degn 1929: 142, translated in *ibid*: 69). The importance of farming during the 18th and 19th centuries is evident in a report stating there "were about 300 substantial farmers, mostly crown tenants, who farmed 2 to 8 marks, but occasionally as much as 30 or even 50 marks"[4] (West 1972:16).

---

[4] A mark (*mørk*) is divided into 16 *gyllin*: A *gyllin* is divided into 20 *skinn*. Then *gyllin* appears to have been a Hanseatic unit that came to the Faroes in the late 15th century. A *mørk* can vary greatly in size. It has been

The 1769 census places the population of the Faroes at 4,773 and another census in 1801 reported an increase to 5,265 (Wylie 1987:70). Wylie remarks: "The essential point is that a population of about 4,500 was evidently near the upper limit that could be supported without increased imports of grain" (*op. cit*). With the production of few trade goods, increased imports of grain did not occur, resulting in little population growth. What is of interest is that while potatoes grow well in the Faroes and are today a home-grown staple food, the Faroese were very slow in adopting potato cultivation and use. While some suggest this is due to the conservative nature of the Faroese to the outside world at this time, Wylie (*op. cit*) suggests a more pragmatic explanation, noting that Faroese farmers have always rejected the plough, which they say is ill-suited to the fragile Faroese soils, and prior to the invention of the motor, required not only as much labour as a spade, but also a horse to pull it, and horses eat hay.

**The Development of Property Rights over Whales**

Joensen suggests the middle of the 16th century as the most likely time the *grindadráp* became the highly organised activity it is today, when "the Faroe Islands declined to the position of a local community without expansive possibilities" (Joensen 1976:6). Joensen uses the term 'local community' to describe a society on the periphery of the power centre, consisting of a small population almost exclusively of land-owning peasant farmers and a landless proletariat who were preoccupied with local matters controlled largely by absentee landlords, thus making whale meat a significant supplement to the Faroese diet and well-being (*op. cit*). Whale meat was a high-quality food source that the Faroese could obtain without being reliant on external economic markets or inputs. This is what Ostrom describes as a time when the working rules or rules in use are set by individuals to organise repetitive activities (Ostrom 1992:19).

Sanderson draws our attention to the writings of the Norwegian Peder Claussøn Friis (1545–1614), whose book titled *A description of Norway and Surrounding Islands* was published in 1632. Sanderson informs us this is "the earliest explicit description of the driving of schools of small whales in the Faroes, and indeed the first recorded use of the element *grind*, to refer to groups of small whales" (Sanderson 1992:50). Friis states he recorded the following account of the Faroes according to a report by a student born in the Faroes named Jacob Dudnesøn:

> There are also various kinds of fish around the coasts there, just as around Norway, although no Trollwhales come there, but large whales [*røerval*, *cf.* modern Faroese *royður*: blue whale] and small whales, 18 or 20 ells long, are sometimes driven in by the Whaledog, when they swim in schools or when they are mating, and when they come in to the bays the inhabitants pursue them with boats and harpoons, and chase them ashore, for where one or two whales precede, all the others follow after them, and it often

---

suggested that a *mørk* of land is what is needed to keep 80 of the old breed of sheep or 40 of the new (Wylie 1987:208)

> happens that a whale takes a boat and its crew a long way on its back, and some boats are capsized, and sometimes a man or more are drowned, and others are saved; some of the whales also escape and go back out to sea.
>
> In the year of our Lord 1587, 300 small whales were harpooned and slaughtered and driven ashore in this one year in the islands, and such has occurred in ancient times, and usually happens every sixth or seventh year, and in thick fog, so that the whales cannot see the land when they come swimming together in large groups, and they call this fog *grind*-fog [*grindamørcke*], that is the kind of fog or mist which leads this school of whales [*hvalsgrinden*] to land.
>
> Anything which is made of wood or iron in the form of a trellis/latticework [*tralleverck*], such as a grill or gate, is called *grind* in Norwegian, and because these whales swim side by side in large groups when they are mating, then that group of whales is called a *hvalsgrind*, as was discussed in the treatise on fish (Storm 1881:431-432 translated in Sanderson 1992:52).

Friis's account of the *grindadráp,* suggests Sanderson, is that "the use of the element *grind* in the context of hunting schools of small whales in the Faroes certainly goes back further than the year 1592 when this particular account was recorded… . Tax accounts for the years 1585, 1586, and 1587 have not survived, so the report cannot be verified by contemporary historical records" (*ibid*: 53).

Sanderson also notes the first existing record of what have been presumed to be pilot whales can be found in the bailiff's account book of 1584 where he refers to the stranding of four small whales. It is now thought that they were not pilot whales but rather bottlenose whales (*Hyperoodon ampullatus*). What is of interest to this research is not so much if they were pilot whales or not, but the bailiff's reference to the division of the catch, another early piece of evidence that property rights to the products of a catch were firmly established as a cultural norm in the 16th century. The bailiff records:

> On St. Hans day in the summer in the midst of a dreadful cold blizzard which prevailed in the Faroes, thanks be to the divine and miraculous providence, 4 small whale fish, called *nyningur*, stranded in Little Dimon [Dïmun]. The dean, Simon Jessøn on Suderø [Suðuroy], collected his majesty's share of the aforementioned whales, which were valued at 4½ *gyllin…* (translated in Sanderson 1992:54)

As a result of this account, the year 1584 is often referred to as the beginning of the Faroese pilot whaling records, which continue today,[5] and from

[5] Faroese whaling statistics, initially found in tax accounts, are perhaps the world's oldest statistics of wildlife use, dating back to 1584 and continue today with the only gap being from 1640 to 1709 (Bloch *et al*. 1990:36)

the bailiffs records, we know that small whales were assessed according to the monetary value of the day, and were divided so that the King subsequently received his proportion of the catch. The hunting of small whales continued, and:

> Surveys of the records for catches of *små hvalfiske* [small whales] shows that from the year 1599 to the mid-seventeenth century schools of small whales 'came in' (*indkom*) on a regular basis. The size and number of the schools taken each year increased from around 1615, when there was often more than one school a year, some numbering several hundred whales (Sanderson 1992:55).

The driving and beaching of whales was, in the 16th century, as it remains today, a significant part of Faroese life and as such was being specifically noted in reports of visitors to the islands, several of whom mention pilot whaling as a distinctive part of Faroese life. These descriptions are important as they provide information on Faroese whaling practices and equipment used in the 17th century. For example, from an account from 1673, by Lucas Debes who came to the Faroes in 1651, we know there was an already well-established and complicated communication system (*grindaboð*) for alerting the attention of the men of the Faroes to the sighting of a school of pilot whales. Debes describes how fires were lit, turf dug and turned, or sheets laid out at particular places to signal that pilot whales had been sighted and where they were. On sighting, the signalmen were then required to row to the location to assist in the drive (*ibid*: 59-60). Highly coordinated activities, such as those described above, ensured that those making the contribution would be ensured some benefit. To maintain this match between contribution and benefit it needed to be clearly established in advance what the benefit would be.

The methods of driving, beaching and killing the whales are also described at length, including the "great crying noise and casting of stones" which occurs in the process of "driving them as fast as they can upon the sands" (Debes 1676:173 translated in *ibid*: 60).

Others, such as Friis, comment on the danger faced by the men in their boats in the course of driving the whales ashore, where they provide a huge economic resource to small geographically-isolated communities. We can see the Faroese view of whales in the 17th century from Friis's description of the Faroes (based on the report of Jacob Dudnesøn), where the appearance of whales is shrouded in mystery with fogs seen to be leading the whales ashore. Emphasis is also given to the chance appearance of a *grind* and its association with a powerful and uncontrollable environment which is then brought closer to the realm of the society and the land and therefore closer to human control. "But as long as they were alive and swimming, schools of pilot whales were, no matter how close to the land, still a part of the 'wild' as opposed to the 'domesticated' sphere of existence" (Sanderson 1992:58).

In 1673 Debes described the system for the division of the catch in some detail. A fairly simple elaboration of medieval practices had been worked out, which took into account both the discoverer of the *grind*, and the church, as well as the hunters and the landowners:

> When all the whales are thus pulled out on to dry land and are counted, the tithe is first taken out; next the finding-whale for the first who saw it; the remainder is divided into two parts, the one part for the people, the other for him who owns the land, whether this be the king, a noble, or a free-hold farmer (Debes 1673:76, translated in Wylie and Margolin 1981:120).

Debes does not tell us who 'the people' are, or who was in charge of the hunt or the distribution. Wylie and Margolin (*ibid*) suggest 'the people' were the hunters, while Joensen (1976:20) notes: "It is apparent from our earliest sources that after the school was dead the [sheriff] assumed leadership for the affair." What these early writings demonstrate is that there was an established and functioning institution for the management of common property resources in operation in the Faroes in the 17th century.

Property rights to whales were reviewed again in the early 18th century, when a review of the whaling laws occurred. The review is detailed in *The Commission Report* (*Kommisionsbetænkningen)* of 1709/10. The Commission Report details how the Danish king increased what were called 'government shares':

> whereupon in the previously described manner the largest whale is separated from the group, which has been killed, as the sighting-whale *(findingshvali)* for the one or ones who first sighted the whales out at sea or in the fjord; then they are all assessed, small and large, by men who have been appointed by the sheriff, and each fish is marked with its number and value. A tenth of the total assessment is set aside; thereafter are set aside five ordinary presentation whales *[foræringshval]* for those who are confined to hospital, for the pastor, for the judge, the sheriff, and the governor; thereafter whatever damage has been done to the boats during the kill, although not always, which according to our most humble and modest opinions nevertheless would only be just; hereafter the whales are divided into two shares (which is otherwise known as the controlled men's share), and thus has been the practice since time immemorial; the inhabitants are of this opinion by law, page 850, article 6, most graciously agrees. The share which is set aside for the land is divided between Your Royal Majesty and the commoners of that place of which there are some, '*after advenant af Marketallene*' (i.e., according to the possession), although at other places Your Royal Majesty retains the whole half-share. The other half-share is divided among the enrolled men, namely each adult man who participates receives a share for himself and his weapons two shares, each hired man ('*dreng*') who can row out to fish one share, and minors who nonetheless can row with an oar 2 or 3 to a share, and each boat

> one share (*Kommisionsbetænkningen* 1709/10:79, translated in *ibid*: 22-23).

The next development in the *grindadráp* occurred during the time of Jens Christian Svabo (1746-1824), one of the first Faroese educated in fields other than theology. His subjects of study were economics and natural history, especially the Faroese economy and the use of natural resources by the Faroese. Svabo's writings in the 19th century were influenced by the Danish Government's inquiries into the reform of Faroese agriculture in order to achieve better economic outcomes and motivated his thinking on the use of pilot whales by the Faroese. Although these economic changes had their genesis outside of the Faroes, we see Svabo actively working as an insider attempting to shape the economic changes so that greater benefits would flow to the Faroese from their use of whales, including the minke whale. However, the use of the minke whale was in conflict with local belief about this particular species. In Faroese, the minke is known as the *sildreki,* the chaser of herrings. It has similar names in other Nordic countries and comes from the old Nordic folk belief that minke were good because they chased the fish toward the fishermen and for this reason they should not be hunted. Svabo considered this belief ridiculous, and argued that instead of protecting them, the Faroese should be catching them as they were in competition with the fishermen (Djurhuss 1976, translated in Sanderson 1992:83).

One of Svabo's main concerns when looking at economic development opportunities associated with the pilot whale hunt, was what he saw as the lack of regulated and standard equipment, especially whale spears (*hvalvákn*), in the hunt. He cited an incident in Vágur (Klaksvík) on Borðoy when a large school avoided capture. At the time this was blamed on the presence of a pregnant woman,[6] but Svabo discounted that superstition, asserting that the real cause was much more likely to have been due to a lack of whale spears (Djurhuss 1976:252, translated in *ibid*: 83).

Svabo, writing in 1779, described the division of the catch. He refers to the '*Findings-Hval'*, (the finder's whale) and to a whale which is given to the poor, and to a number of 'presentation whales' that were given to certain prominent persons in addition to widows and the poor. For example, in the 18th century, the asylum in Arge received one whale for the benefit of its patients. In 1767, after the asylum closed, it was proposed that the presentation whale be used to establish a poor fund (Bjørk 1963:241, translated in Joensen 1976:25). Svabo details a division, which occurred in 1747, where:

1. a tenth part is set aside
2. to the injured, 5 skinns
3. to the poor, 10 skinns
4. to the assessors, 1 florin 5 skinns
5. for the report to the sheriff, 6 skinns
6. the watch of 10 men, 12 skinns

[6] According to an old Faroese superstition, having pregnant women or priests witness a *grind* brought bad luck.

7. boat damages, 2 florins 7 skinns
8. the land ½ and the hunters ½ [7]

(Svabo 1959:257 translated in Joensen 1976:23)

Svabo also criticised the traditional apportionment of half the catch to the landowners, which he compares with the division of a catch of fish, calling for a change in these laws to allow a greater share of the whale catch allotted to the participants in the drive and kill (Djurhuus 1976:259, translated in Sanderson 1992:83). Here we see Svabo attempting to redirect resources flowing from the periphery to the core, where they would benefit the Faroese rather than absentee landlords:

> The impractical distribution of the whales ought to be changed. The land on which the kill takes place receives a half-share; no wonder, then, that each wants very much to have the school driven into a bay where he has land and home. One argues and complains vigorously about this instead of paying attention to the hunt, and it is certainly not always the best advice which is followed. This distribution system was probably based on the old Norwegian law… of which Christian V's Norwegian Law 5, 12, 6 seems to have been consulted: 'if a whale is shot or driven ashore, the one who shot it owns half-share and he who owns land ½', but this law's disposition seems unnatural as it is outmoded. That which the fisherman otherwise catches is his own, but when he here makes use of another's bay to drive in his whales, he must give more than half of it in rent since recompense for damages is only taken from his share… if one wants to give the landowner a share of the catch for the sake of time worn justice, ¼ or rather one-sixth in my opinion, would be too much rather than too little (Svabo 1779:52, translated in Joensen 1976:23).

The landowner share dates back to the *Seyðabrævið* in 1298, where three-quarters of the catch was paid as rent to the landowner, with only one-quarter going to the hunters and associated community. Joensen describes how the land tenure changes at the close of the 17$^{th}$ century impacted the division. The landowner could be the king, a noble, or a commoner:

> Real property on the Faroes has been in part privately owned land or commoners' land and in part owned by foreigners. The families Rosenkrads and Benkestok from Denmark proper owned much property on the Faroes. At the close of the 17$^{th}$ century these holdings were sold. The crown lands were held by the 'King's tenant farmers.' Land which crown officials held *ex officio* also

---

[7] The sum total of the initial payments is taken from the grand total and the remainder is declared *partahvalur*. It is from the *partahvalur* that ½ is divided between the land and the hunters (*see* Williamson 1945:126).

had full rights to *jarðarhvalur* [the land whale] while there were other rules for crown-lands (Joensen 1976:24).

Around this time, several other voices joined in the cry demanding that landowners' rights be abolished. Sheriff Hammershaimb expressed a similar view in a document drafted in connection with the Executive Orders on whaling of 1807:

> It is truly severe that often those, who with night-long, lengthy journeys in the most inclement weather, dank and damp, not infrequently do not receive very much in exchange to take home to their tables while he who stands on shore and watches, not to mention that distant absentee landowner, gather in the riches, even though they only own that bit of land and have not the least to do with the hunt (Bjørk 1963:246, translated in *op. cit*).

Svabo relates that some people tried to avoid conflict by holding the shoreline '*i Fællig*' (in common) so that the landowners in the area together owned a whole stretch of coast (Svabo 1959:254, translated in *ibid*: 25).

There was a campaign to abolish the *jarðarhvalur* (land whale). This was debated in the Faroese legislature and in the Danish national legislature body. As a result the *jarðarhvalur* was not abolished at once, but reduced to one-quarter of the pilot whale catch. The other half-share was taken from the landowners without compensation, but when it was finally rescinded in 1934, the landowners received recompense for the expropriation of their prerogatives (*ibid*). There were other problems occurring at this time:

> Thievery, stealing, and disorder rules here, and the *sysselmænd* are always too weak to maintain order as they wish. This disorder became entrenched too long ago for it to be possible to change, for it is not considered thievery but rather pluck or manliness to steal whales (Svabo 1959:256, translated in *op. cit*).

Landt also commented on the problems with the *grindadráp* at this time, noting.

> the distribution notwithstanding, a great deal of thievery, dishonesty, and wilfulness is practiced so that some seize and take away twice as much as they deserve and others on the other hand go away with little or nothing. Also, another old abuse allows everybody to cut from the whales' fins the customary oar-bands for the boats so that some are able to appropriate this advantage, but most receive nothing although some of these in turn do not want to be deprived of their rights completely and so slash further into the whales and thus mishandle them at the expense of those to whose lot they fall. It is therefore highly desirable that a better system be established for this distribution; that man who comes up with the

best proposal will do the country a great service (Landt 1965:226, translated in *op. cit*).

There was growing concern about regulations introduced to resolve conflict and provide more transparency: "this hunt… cries for a whaling law which ought to control everything having to do with the hunt and which is based on the advice and experience of the oldest and wisest men regarding the drive ashore and the kill with respect to each whale-bay" (Svabo 1959:257, translated in *ibid*: 39).

During this period only two design principles discussed earlier are evident, *viz.,* clearly-defined boundaries, and a match between participants' contribution to the *grindadráp* and benefits received from it. However, at that time, both are rudimentary and still evolving; there is no evidence of collective choice arrangements being in place, allowing most individuals affected by operational rules to participate in modifying them. During this period, the *grindadráp* is monitored by the sheriff, however, this form of monitoring is top-down with the users having little influence or involvement. In the early literature there is no detailed evidence of graduated sanctions and none relating to conflict resolution. With only two rudimentary design principles evident, the *grindadráp* during this period could not be considered robust.

## The Third Period: 1800–2000

The early 19th century is seen as the beginning of modern Faroese history, with swift social change accompanied by the emergence of economic growth in the export fishery. The end of the 19th century saw the emergence of a written medium of expression in Faroese, created by V.U. Hammershaib, and adopted as the standard orthography of the language, which, it is suggested, was "the first decisive step in the reformulation of Faroese culture" through which the Faroese were able to write their own histories and systematically collect and publish their own folklore (Wylie 1995:53). Faroese authors rapidly developed a modern literature of poetry and novels, and "political separatism followed a pace behind, dependent on successful demonstrations of cultural distinction" (*op. cit*).

Denmark lost Norway, through the Treaty of Kiel in 1814, but retained Greenland, Iceland, and the Faroes. The old Faroese–Norwegian law was kept in effect, and the Faroes were governed as a Danish county. The *Løgting* was reconstituted in 1851-52 as a consultative body along the lines of other Danish assemblies, and economic reforms were designed to make the Faroes a more prosperous part of the Danish realm. Emphasis was given to developing commercial fishing, based on expanding the traditional subsistence fishery from open boats so that this new industry would slowly replace the older economy based on woollens (*ibid*: 70-71).

Initially the population of the Faroes had remained constant around 4,400, but by 1801 it had grown to 5,265 and reached 11,220 by 1880 and 15,230 by 1901 (Wylie 1987:113-114).

The early 19th century saw the emancipation of the farm servants, but unlike Britain and Europe, where these workers moved to larger industrial regions to be employed as factory workers, the Faroese labourers were able to utilise the sea and its resources and also gain access to small plots of land (*trøð*) for their own use for cultivating potatoes or keeping a small number of sheep or a cow or two for milk. The agricultural products along with the products from pilot whaling and fowling were consumed by the families themselves, while the surplus from the fisheries were sold for cash or, more frequently used to exchange for the imported products of coffee, alcohol, salt, grain, timber and other exportable commodities (Høgnesen and Apostle 2002). As Faroese farm labourers moved into the emerging individualistic and profit-focussed commercial fishing industry, most were able to remain in the small villages and therefore remain in what Bulliet (2005:3) described as a domestic society where people view as commonplace and acceptable the killing of animals for food purposes.

In 1837, Christian Pløyen, a Dane, was appointed governor; Pløyen was regarded as quite an innovator for his time. He, along with some Faroese men, went on a fact-finding trip to the Shetlands to look at their emerging fishing economy, particularly the use of long-lining, and to see if the ideas used there could be employed in the Faroes. The period was marked by increasing reform, *e.g.,* a savings bank was established in 1832 to absorb the increasing private savings. During the 1830s, the trade monopoly liberalised its policy on purchasing fish, and from 1845 it began to buy fresh fish, something that had not occurred in the past. The trade monopoly, primarily based in the capital Tórshavn in the early 19$^{th}$ century, began to open branches in peripheral areas of the Faroes in 1834, 1837 and 1838, including in Tvøroyri on Suðuroy, Klaksvík on Borðoy and Vestmanna on North Streymoy (Høgnesen and Apostle 2002).

The trade monopoly's change in policy in regard to buying fresh fish, further facilitated the development of the fishing industry as the fishermen were now exempt from having to process the fish, and could concentrate on the business and activity of fishing. This small change laid the foundation for transforming the Faroese economy from one based on agriculture with few tradable goods and with severely restricted trade controlled by the monopoly, to one based on fishing and the capitalist market economy. When the trade monopoly was finally abolished in January 1856, its buildings were primarily bought by Danish and German merchants, with other smaller merchants affiliated with the stronger merchants opening businesses around the islands. The merchants were not merely grocers: many were bankers, fish buyers, sellers, and processors, further stimulating the growing marine based economy. The end of the trade monopoly allowed Faroese who had worked for the monopoly to utilise their business skills to open their own businesses around the islands. In 1858, J. Mortensen from Øravík obtained a licence to trade, buying knitted socks, woollen sweaters (jerseys) and fish in exchange for the most necessary commodities. Over the years his business slowly grew to include 20 branches around the Faroes as well as about 30 fishing and cargo ships. The business was later taken over by his son who continued to expand the business (*op. cit*).

As well as expanding commercial fishing opportunities, the Faroese were also looking at expanding whaling to other species for commercial trade and to provide a domestic meat supply to feed the growing population. For a limited time, the Faroese engaged in commercial whaling, which was initiated in the Faroes by the Norwegians in 1894. At the beginning of the 20$^{th}$ century there were seven whaling stations operating on the Faroes with a fleet of 17 whaling boats. The catch of whales in 1909 was recorded as 773 large whales. As was the case elsewhere, the economics of commercial whaling in the Faroes fluctuated greatly with world whale-oil prices.

During this time, the censuses show the proportion of Faroese population supported by agriculture fell from over 80 percent to under 30 percent, although the absolute number remained nearly constant, at around 4,400. The growing population was supported by the new and burgeoning commercial fishery and associated enterprises (Wylie 1987:113). The last quarter of the 19th century was also marked by the drop in the prices paid for agricultural and knitted products and the rise in prices paid for fish, adding greater incentives for people to enter the marine economy.

With the growing economy came an increase in population, an increase which placed increased pressure on land use. For example, in Fuglafjørður by the middle of the 19th century, four-fifths of the land in the community was in private hands, with most of it being owned by people living outside the village. "The lots of land got smaller and smaller by partition on inheritance; this in turn, forced the small-holders to intensify their fishing activity" (Høgnesen and Apostle 2002:79).

Despite unsuccessful attempts in the mid-19th century to establish a Faroese deep-sea fishery, by the 1870s it had become an established business and the long-term foundation of the Faroese economy. Initially, Faroese fishing boats were small coastal craft, but that changed when the introduction of steam vessels in Britain resulted in larger sailing vessels (sloops) becoming available on the market, whereupon they were purchased by the Faroese.

Capital needed to enter this business came from the fishermen themselves and from businessmen who were able to purchase the relatively cheap British sloops. By the beginning of the 20th century, the economy was driven by the fast-developing fishing industry. Distant fisheries began to be utilised by Faroese fishermen, with larger crews of 15 to 20 men operating for extended periods of time away from the Faroes. From early March to late May the sloops first operated in Faroese waters and later, during the first decade of the 20th century they began to operate in waters south of Iceland. Between fishing trips, fishermen would plant their potatoes in early June and then go to sea again for fishing trips north and east of Iceland until the end of September. When the vessels returned, they were de-rigged for winter (*op. cit*).

The increase in fishing activity and growing export markets provided an opportunity for the entry of women into the fishing industry through jobs in processing plants that were established to process the larger catches. This influenced internal migration away from small villages to developing regional centres such as Runavík, Sørvágur, and Vestamanna (*op. cit*).

By 1921, diversification of industry began, as the growing fishing industry could not absorb all the new entrants. Developing secondary industries included light industry, trade, services and crafts (many of which underpinned the fishing industry), while agriculture continued to decline. These new developments, in common with the past fortunes of the Faroes, were dependent on foreign trade and relations with core centres of economic power. Faroese fishers, like Norwegian fishers, had long-established markets in the Catholic countries in the Mediterranean region. However, the anti-fascist boycott of Mediterranean countries, beginning in 1936 and lasting until the end of World War II, cut off access to those traditional Faroese fish markets. The collapse of these markets totally altered the structure of the Faroese fisheries and thus the occupational structure of the industry, causing widespread unemployment, especially for women who processed the salt-fish (*op. cit*).

Since the mid-19th century, Faroese politics had to deal with the emerging phenomena of modern business and the market economy. The political process was very much influenced by the complex structure of interests inherent in the conflict between the declining position of 'crown' farmers and the emerging class of businessmen and ship owners. From the perspective of profitability, the outbreak of the World War II came as a relief. The war between Britain and Germany created an unmet demand for fish in British markets, an opportunity that Faroese businesses moved to fill. Prices for fresh fish were very high on British markets, although the risk of being attacked by German planes was very high, with many Faroese men losing their lives while transporting fish to Britain (*ibid*: 83).

World War II also brought a significant and lasting change to the Faroese political system. While Germany occupied Denmark, Britain occupied the Faroes and this brief separation from Denmark increased the republican sentiment amongst many Faroese people—so much so that a referendum on independence was held in 1946. While a slim plurality voted for complete independence, the referendum was annulled. However, in 1948 a Home Rule Law gave the Faroes a new status (Wylie 1995).

Commercial whaling in the 20th century paused during the two world wars, and after World War II it began again on a much smaller scale, with only two shore stations operating in the Faroes. The total catch in 1950 was just 432 whales. As whale-oil prices fell, along with the stocks of the great whales, the industry was no longer economically sustainable and this saw one whaling station close in 1954 and the second in 1958. The 1950-51 statistics indicate that whale products from the shore stations made up only 8 percent of Faroese exports. While commercial whaling in the Faroes contributed to local employment, its most important contribution to Faroese society was a source of cheap healthy meat. It was this desire for a domestic supply of whale meat that saw a small-scale whaling operation established again from 1962 to 1968, with one vessel operating from a single landing station. The annual take of whales was less than 20; fin whale meat sold well domestically, while the oil was exported. This operation ceased in 1968, but another started in 1974 aimed to catch minke whales for domestic consumption. This coincided with a period of very poor pilot whale catches around the islands when there was a strong need for sources of meat (Olafsson 1990).

All these catches of 'great whales' by the Faroese were carried out under the jurisdiction of the IWC. In 1976, the IWC set a number of precautionary quotas for whale stocks, 'freezing' the catches at their actual levels, including a zero quota for commercial catches of the so called 'South Norway–Faroe' fin whale stock, where recent catches had been zero. The Danish Government considered the zero quota set by the IWC to be applicable to the small Faroese operation, a view that the Faroese Home Rule Government did not share. This saw disagreement between the Danish and Faroese Governments concerning the division of powers in relation to whaling. Since the introduction of Home Rule in 1948, the Faroese have had the power to regulate marine resources in the Faroese fisheries zone, which prior to 1977 extended to the 12-mile limit; fisheries beyond this limit, at that time considered international waters, being the responsibility of Denmark. This changed in 1977 with the introduction of the 200-mile limit, and Faroese authorities claimed the power to regulate all living marine resources, including all whales, within this area. The Danish Government at first did not agree, but the dispute was finally resolved with the Faroese Home Rule Government introducing amendments to the whaling legislation in 1984 (*op. cit*) that read:

> All whales (cetacea) are protected from whaling, regardless of the method used.
> Point 2. In special cases the Home Government, in conjunction with the Fisheries Committee has authority to exclude certain species from this protection (*Act of Faroese Parliament on Amendments to the Whaling Act* 1986).

This effectively saw the Faroese Home Rule Government now holding responsibility for all whales within its 200-mile economic zone on the same basis as its responsibilities for fisheries and other marine resources.

Home Rule meant that the Faroes was no longer a Danish county. The Faroes became internally self-governing in some matters, including taxation, customs, regulations and economic affairs; other matters were controlled jointly by the *Løgting* and the Danish Government, which is represented by a High Commissioner residing in Tórshavn. Currency, defence and foreign affairs remained the exclusive domain of the Danish State, with the Danish Supreme Court as the court of appeal for the Faroese court. The Faroes also elect two members to the Danish parliament. Wylie notes:

> Separatist goals were realised in 1948 in the Home Rule Law by which the Danish government granted the Faroese internal self-government on, precisely, the cultural, political, historical, and natural grounds wherein Scandinavians recognise their distinctiveness, or as the preamble of the Home Rule Laws puts it, in recognition of the special position which the Faroe Islands occupy with the realm in national, historic and geographic respects (Wylie 1995:53).

The *Løgting*, the Faroese legislature, has its representative roots in the early settlement of the islands. Its members choose among themselves an executive committee, called the *Landsstýri,* with the head person of the executive committee, in effect, the Faroese Prime Minister. With the economy dependent on fishing, the *Løgting* decided that the Faroes would not join the European Economic Community (EEC) when Denmark joined in the 1970s. The EEC fishing policy was unacceptable to the Faroese as it was based on free access for member countries to all Faroese waters outside the 12-mile limit. The

Faroes could not benefit from the diversified industry of the EEC countries, and it was thought that more rights and benefits would be lost than gained through membership, a view that still predominates today (Holm and Mortensen 2002).

**Codification of Custom in the Recent Past**

As discussed above, during the colonial period, the institution of the *grindadráp* remained weak for a number of reasons. For example, the first two design principles, *viz.* clearly defined boundaries and a match between contribution and benefits, were still rudimentary in their development. There was no community-based or local-level monitoring, which would see monitors accountable to the users, and no graduated sanctions or conflict-resolution processes. However, during the recent historic period there is evidence of positive changes occurring.

By the beginning of the 19th century, a number of proposals began to be presented in the Faroese legislature to establish greater cooperation and order on the whale hunts. On June 14 1807, Governor Løbner and Sheriff Hammershaimb sent a draft document outlining proposed whaling regulations to the Danish Government in Copenhagen. This draft stipulated extremely harsh punishments for those who disregarded the regulations, e.g., "Anyone who acquires his portion of the catch through violence or coercion was to be punished with 4 years' labour in a house of correction, and anyone who used a harpoon during the whale drive was to be given a similar sentence" (Bjørk 1963:188, translated in Joensen 1976:39).

A subsequent version sent to the Office of the Exchequer in 1807 shows how chaotic the situation appeared to the authorities:

> Those severe punishments here and there among the accompanying regulations could perhaps under other circumstances appear too harsh, but if evil is to be driven out with the rod, then it certainly is to be done here in the Faroes where there is no other solution than with strong and exemplary measures, which are to be applied to the utmost on each and everyone who transgresses them, crush the disorder so prevalent in the past; a public example will, according to our knowledge of the Faroese character, effect more than fines which they with ease even agree to take from the common fund (*op. cit*).

The proposed harsh punishments never came into force, and more drafts were debated until finally, in 1832, it became Governor Tillisch's responsibility to see the first Faroese Executive Orders on Whaling adopted (Joensen 1976:39-40). Since that time, every detail of the hunt has been carefully defined in the Executive Orders, which over time have adapted to new situations. What we see in the Executive Orders on Whaling is the recognition of the old institutionalised customs of the pilot whale hunt together with new ordinances when and where these customs have proved insufficient. In the view of Joensen (*op. cit*), the whale fishery came to be considered by the Danish Government as so important for the Faroes that legislated regulation was judged necessary to keep intact this economic benefit to the Faroese. The institution of the pilot whale hunt, which earlier was based largely on tradition, was now to be integrated formally into the society's legal structure.

Another factor influencing the development of the Executive Orders was that toward the end of the 18th century there was a period when there were what can only be described as 'grindless' years, when there were no whales taken at all. No one can say for certain why this occurred, but it had an impact on the way the Faroese viewed the institution of the pilot whale drive. The Faroese viewed these 'grindless' years as a danger to their long-term use of pilot whales, the danger being the loss of the knowledge and technology utilised in the drives. In periods when there were no *grind*, they believed a whole generation may pass without ever being involved in a hunt, and the knowledge may die with those who had that experience and knowledge. As pilot whaling is opportunistic, people must always be prepared, with the appropriate technology to act quickly and cooperatively to be successful. When people are ill-prepared, such as by lacking the necessary technology, the hunt will be a failure and no food will be secured. It was the view of Svabo and others that something had to be done to keep the knowledge alive. One means to achieve this was to codify the customs of the hunt in the form of Executive Orders that could be amended from time to time as needed. The Executive Orders would also stipulate what technology was required, who would control the hunt, and the how the distribution of the products was to be effected. The development of Executive Orders would also provide an opportunity to review the distribution practices that Svabo had criticised (Joensen 1976:26-28). It was envisaged that codifying customary practice and drawing previous legal decisions together into one law would be seen, on an official level, as a means of food production which could be controlled and made economically viable (Sanderson 1992:86).

What also stimulated the development of the Executive Orders was the high number of whales that were being taken in the early 19th century. The Faroes at that time was experiencing a peak in the number of whales, a situation that Hoydal and Lastein (1993) suggested may be related to environmental factors. With large numbers of whales being taken regularly in Faroese waters over a number of years, the commercialisation of whale oil for trade in the developing market economy was pursued to further support the growing population. This focus also required that opportunities to take whales were not missed through a lack of organisation or equipment.

The development of the Executive Orders on Whaling is especially significant if we consider that, at the time, whale oil possessed significant value as an export commodity, representing 20 percent of the value of all exported goods (West 1972). The formalisation of the pilot whale drive can be seen as an attempt to get as much benefit from the resource base as possible, and to exploit it to a greater degree than had occurred before. At this time the balance in the use of pilot whales shifted from the domestic economy toward a market economy.

An example of external economic forces asserting an influence on the introduction of the Executive Orders on Whaling can be seen with Pløyen and his drive for economic efficiency in the hunting of pilot whales, so that the oil could be used for trade in the market economy and the meat remaining in the subsistence economy. Pløyen has been described as a great aficionado of the *grindadráp* (Wylie and Margolin 1981:122) and one of the things he was involved in during his time as bailiff was the establishment of a net in Vestmanna to be used in the pilot whale drives with the aim of making whaling more efficient. Whaling in open row-boats, given the tides, currents and bad weather conditions, could lessen the chances of a successful catch. This situation ensured that people were always looking for ways to make the *grindadráp* more productive. For example, the whaling bay at Vestmanna (Vestmanhavn) was considered, before 1843, one of the inferior whaling bays; Müller (1882:9) described how the sloping bottom of the bay was considered too narrow and at the side of the bay there was a '*marbakki,*' (a suddenly sloping bottom), which prevents the beaching of the *grind*. Müller also described how in 1841 a *grind* had been driven into the harbour where it had been for a few days without having been taken successfully:

> The harbour is wide, quite a mile long, and it has at the mouth a sand-bar upon which there is three fathoms of water. This bar has a sloping level bottom outside, but inside the bottom suddenly slopes, the result of which is that the *grind* can be easily driven in over it, and finds difficulty to get out again. For this reason a grind can be kept in Vestmanhavn for days (*ibid*: 9-10).

To keep the whales in Vestmanhavn, people tied bundles of straw to a long rope with stones attached, which held the bundles of straw just below the surface to act as barrier which, on being drawn toward the shore, moved the whales closer to shallow water where they could be speared. This method ensured 30 whales were taken at that time. Following that successful outcome, a whale net was constructed from old fishing lines, and although it was a fragile tool the people managed to take 36 whales. Müller described the further development of this innovation:

> When Governor Ployen and Sheriff Lunddahl who took great interest in the whale-fishing, and never neglected to be present, saw this astonishing result, they adopted the pastor's plan of a net at Vestmanhavn, and caused one to be made. This net, which was manufactured by the intelligent blacksmith Peder Fr. Jacobsen... according to the pastor's sketch, consisted of nine yarn ropes, 200 fathoms long, and 8 fathoms high, with meshes of 6 inches square, lead sinkers on the bottom rope, and buoyant at the top, with fifty oak barrels for floats, and was ready for use in the spring 1843. It was fitted up for trawling after the grind, so that it was forced in upon the level bottom and prevented from striking against the 'Marbakke.' The expense amounted to about 100 pounds. A store was built for keeping it in.
>
> It was used for the first time on Whitsunday morning, the 4th June 1843. The draught was successful above all expectation, as 234 animals were killed with only one attack, not a single whale escaping or getting outside the net. This successful result was followed the same year by two catches less successful, as the herds

> broke through the net; but we succeeded nevertheless, after hard work, the net being very heavy, to replace it outside the flocks, and catch both these grinds in the course of two days. The deficiency of the net arose partly from want of sufficient sinking weight to prevent the grind from lifting it from the bottom and swimming out underneath; partly also from the fact that when the bottom line was made heavier, and the grind was not able to lift it, it went over it, although all the boats that could be spared from spearing were fastened to the floating rope. The boats were obliged to allow passage over, or to sink themselves, which happened a few times, although no damage was sustained by the crews.
>
> This deficiency was not easy to mend, as more weight made it impossible to move the net and better floating could not be applied. We were consequently forced to be pleased with the net as it was. Later grinds were partly as successful as the first one, but in 1858 one grind was lost, because the net could not be put out again after the grind had broke through it, as bad weather came on. As the net by this time required general repair, it was resolved to have a new one made, smaller of 100 fathoms length, which was easier to handle, and to use the old one to bar the grind upon a smaller portion of the harbour, by which means much more time would be saved when the grind broke through the smaller net, as it had not a fourth part of the harbour to move in. This plan is best suited to the purpose, and not a single fish has escaped since it was adopted. As soon as the grind is barred by the large net, the smaller one is set from the bottom along the north shore of the harbour. When the grind is to be attacked—which may most appropriately be performed in rising water, as the current from the river Foraa has detained many a catch, driving as it does the blood too soon into the net—it is driven in between the west end of the net and the shore, the end of the net is hauled on shore, and the grind is then locked at the same time as the attack is made. When this is done in good order, a grind will be caught in Vestmanhavn as easily as in the best whale *voe* [bay, narrow inlet].
>
> From 1843 to January 1878 (35 years), there were caught with this implement 6,030 whales, representing in value £20,100, while on the contrary, from 1584 to 1843 (260 years), only 2,169 were caught, notwithstanding that large herds were more frequent in Vestmanhavn than in the thirty-five years in question—here which, after many day's trouble and enormous exertions, loss of tools etc., were lost (*op. cit*).

While interest grew in trading whale oil in the market economy, whales remained important to the non-market economy. Williamson describes the distribution of a *grind* at Miðvagur in 1894 where a total of 186 pilot whales, valued at 1,300 *skinn,* were taken. What is of interest to this discussion is the percentages of the products of the hunt taken for various community purposes. First, a tithe of 10 percent was taken, followed by the other traditional divisions of the hunt including the *finningafiskur* (finder's whale) and one percent to the Poor Peoples' Fund and another one percent for the schools. From the remaining 980 *skinn,* 25 percent went to satisfy the claims of the landowners (*jarðahvalur*) with the rest being divided up as *partahvalur* (which is the share remaining after the total of the initial payments is deducted from the grand total). The tithe that was taken was split four ways between the church at the *grindapláss* (whale bay and the village(s) near by), the Danish Government, state endowment funds for the church, and the local pastor. The amount that was taken by the landowners was split two ways according to the type of their landholdings (Williamson 1945: 27).

In brief, the distribution consists of first deducting a row of fixed costs and then dividing the remainder among the residents in the whaling district, if the school of whales is over a certain size; otherwise, half the catch is divided among those who have taken part in the hunt, and the remainder is sold, and the proceeds deposited in the islands' public treasury. This latter course of action, however, occurs only when the school of whales is so small that it does not pay to distribute it. The tithe was discontinued in 1907 (Joensen 1976:26).

With the Executive Orders came the introduction of numbering the whales consecutively so that each man would now know from which whale he was entitled to take his share. This number was cut into the head of the whale in Arabic numerals (Fig. 12) and the *skinn* value cut into the pectoral fin in Roman numerals (Fig. 14) (*ibid*: 25). Consecutively numbering the whales was introduced to stop disputes and further clarify from which whale a user was to withdraw his benefits (meat and blubber).

In the final draft of the Executive Orders, it became established that the poor fund was to receive one percent of the catch and the public educational system one percent. After the Executive Orders on Whaling of 1832, the Crown official did not receive a share of the total catch, but rather a share of that which fell to the crown-lands. When the Executive Orders were revised in 1857, the crown-officials' privileges were taken away completely, but the poor fund and the educational system each were allotted one percent until 1939 when this arrangement was abolished (Bjørk 1963:242, translated in Joensen 1976:25).

Throughout the history of the *grindadráp*, there have been successive adaptations made by the Faroese, attempting to balance what Weinstein (2000:381) refers to as the "match between contribution and benefits" (design principle two), allowing participants to believe that they receive an equitable benefit from their individual effort. Throughout the colonial period of Faroese history there was inequity in the distribution of the resources. This reflected the political and economic relations between the core centres of economic and political power and the peripheral areas from where they drew resources. Despite the inequity of the distribution during the colonial period, Faroese people continued to participate in the *grindadráp* as there were almost no other opportunities to access highly valued meat and blubber. The provision of this locally produced food provided the geographically-isolated Faroese communities some degree of food security, and over time, adaptations to the *grindadráp,* especially the rules of distribution, resulted in the more equitable distribution of these foods to the Faroese people.

## The Fourth Period: 2000–2007

In 1970, the population of the Faroes had reached 38,600, peaking at 47,838 in 1989, and declining to 43,393 in 1996 through migration in response to an economic downturn. At the beginning of the 21st century, the population of the Faroes had reached *ca.* 48,000 people, with 18,684 living in Tórshavn, 4,846 in Klaksvík, and the remainder of the population living in small villages spread across 17 of the 18 islands. Life expectancy is 74.1 years for males and 80.8 years for females, and the fertility rate is higher than any other Nordic country at 2.62 (*Hagstova Føroya* 2000).

The modern economy of the Faroes is highly dependent on fisheries and fish products traded in international markets. The great decline in the importance of agriculture and woollen products since the historic past is clearly seen in the 2000 statistics of wage expenditures by industries, which shows agriculture at zero percent (*op. cit*). The dependence on the sea is clearly seen when comparing the size of the land-base of 1,400 km$^2$ to the territorial waters of 274,000 km$^2$. The highly organised and technically sophisticated fishing industry has been productive enough since the end of World War II to support one of the highest standards of living in Northern Europe (*Hagstova Føroya* 2000), with an unemployment rate immediately prior to the 2008 global recession around 4 percent (Tyril 2006).

Thus fishing is by far the major industry in the Faroes, accounting for some 95 percent of commodity exports (exclusive of services and intellectual property); much of the total catch of 261.3 million tonnes is caught in Faroese territorial waters, although the Faroese fishing fleet ranges into other waters (*op. cit*). According to government statistics, the export value of fishery products is Danish Kroner (DKK) 3,668 million;[8] total value of goods imported to the Faroes at that time was DKK 3,738 million. Most Faroese exports go to the European Union, with 78 percent going to the U.K. and 19 percent to Denmark; Spain continues to import Faroese dried fish. Faroese imports come predominantly from Denmark (33 percent) and Norway (18 percent) followed by the U.K. (5 percent) and Iceland (4 percent). It is expected that trade with Iceland will increase following the two countries recently signing a bilateral free trade agreement. Imports also come from New Zealand (for lamb) and Brazil (for beef) (*op. cit*).

The subsistence economy of whaling and wildfowling remain important, especially when "the economic value of pilot whales is measured against the economic and environmental costs of importing food, equivalent to some 30 percent of all meat produced locally in the Faroes" (*Løgmansskrivstovan* 2006).

While the fishing industry is the primary industry in the Faroes, it only accounted for 19.6 percent of total wages paid in 2004. Local and national government is the major source of wages in the Faroes, and accounts for 34 percent of the total. The service industry accounts for another 10 percent and the construction industry 6.5 percent (Tyril 2006). The average annual wage in 2004

---

[8] DKK4.9 = US$1 (in November 2009)

was DKK 210,000 for men and DKK 140,000 for women; the inequity problem is recognized. In 2005, income tax revenues amounted to DKK 6,981 million, paid by 29,960 taxpayers (*ibid*). The discovery of oil in 2001 within the Exclusive Economic Zone (EEZ) may further diversify the Faroese economy; although exploratory drilling is ongoing, the commercial viability of the deposits is not yet confirmed.

The Faroese language remains strong, with some 149 books published in Faroese in 2004, including 67 translations of other works. This is one of the higher number of books published per capita in the world: three books per 1,000 people in the Faroes, compared to two books per 1,000 people in the U.K. and 0.5 books per 1,000 people in the U.S. (*ibid*).

The Faroes has a very low crime rate and little funding is allocated to 'public order safety.' However, government spending on social welfare amounts to DKK 1,865 million out of a total budget of DKK 5,328 million (*ibid*). The current national economic situation has been characterized in the following manner:

> The Faroese economy has moved strongly in a post-industrial direction. Around 60 percent of the country's firms are located in trades and services. Given the country's history and location, it is also no surprise that the primary industry, largely fishing, remains numerically stronger (but not in terms of overall assets) than manufacturing. Further dominance of Tórshavn in the national economy is well understood. It has at least 3 percent of the country's firms based there, and 64 percent of its assets. Tórshavn clearly dominates the tertiary sector numerically; when one takes account of mean values, Torshavn controls over 60 percent of the country's manufacturing assets. The rest of the country does the fishing... . [F]or analytic purposes, it is possible to divide the country into three main groups; the 'centre,' the 'inner periphery' and the 'outer periphery.' The centre is the Tórshavn area, while the inner periphery consists of those municipalities that have over 1,000 residents. The outer periphery consists of municipalities with fewer than 1,000 people. There are only eight municipalities on the inner periphery, and 41 on the outer. Not surprisingly, the outer periphery is quite heterogeneous, comprising everything from traditional remote islands, like Fugloy, Svinoy and Mykines, as well as disadvantaged smaller areas, ranging from Sumba to Gjøgv. However, the situation is even more complicated, as small municipalities near Tórshavn are correctly associated with this group. They include places like Hestur, Kirkjuabour and Nolsoy (Hovgaard and Apostle 2002:23).

The 'outer periphery' communities collectively account for over 25 percent of Faroese business, but their mean value, save in the primary sector (fishing), is lower, due to both size and capital limitations. The 'inner periphery' contains eight separate municipalities. For example Klaksvík is by some distance, with over 4,700 residents, the largest municipality outside of Tórshavn.

It remains a port community, with many small fishing businesses in the primary and the secondary sectors and is also the service centre for the northern islands. Runavík and the adjacent municipality of Nes, is the economic core of southern Eysturoy, and represents the major economic centre outside of Tórshavn. Fuglafjørður is the core municipality on the northern part of Eysturoy. Tvøroyri and Vágur are the two centres on the southern island of Suduroy. Vestmanna, near Tórshavn, has a major hydroelectric facility and considerable aquaculture activity, but also a marked dependence on the capital for trades and services. Miðvagur, with the international airport, has considerable economic autonomy (*op. cit*).

While Danish and English are taught in schools, Faroese is the first language of instruction as it is the language of the state, media, theatre, and everyday life. The Faroes remain outside of the European Union and Faroese citizens can choose to travel on either a Danish or a Faroese passport. Many Faroese choose the latter, with much debate (in 2007) taking place on sovereignty, with several political parties established to pursue this goal. Six parties contested the 2001 elections, *viz*:

- *Tjóðveldisflokkur* (Republican Party)
- *Fólkaflokkur* (People's Party)
- *Javnaðarflokkur* (Social Democrats)
- *Sanbansðarflokkur* (Unionist)
- *Sjátvstýrisflokkur* (Independence Party)
- *Miðflokkur* (Centre Party)

Despite this large number of political parties, all share common support for the sustainable use of their marine resources, especially pilot whales. Whaling has never been a divisive issue within domestic politics, where whaling is seen as one small part of Faroese life. One party, *Tjóðveldisflokkur* has proposed enshrining Faroese rights to whaling in a Constitution once the Faroes gain independence from Denmark.

As part of the Danish Kingdom, the Faroes are bound by international obligations entered into by Denmark. However, the Faroese authorities may influence decision-making in international matters of Faroese concern through consultations with Danish authorities and participation in Danish delegations and, on a case-by-case basis, may negotiate in their own right, assisted by the Danish Foreign Ministry. However, in relation to the International Whaling Commission, the whole Kingdom of Denmark (Denmark, Greenland and the Faroes) is included in Denmark's membership, even though Denmark itself does not whale. Legislation on whaling is made exclusively by the Faroese Government, with due respect to international obligations, including any binding IWC rules that legally apply. Legislation on related matters, such as animal welfare and hygiene, also comes under the authority of the Faroese Government.

Faroese government officials are engaged in establishing new relations with foreign countries, cooperation with relevant UN bodies, membership in international organisations, new free trade agreements, tax treaties and international collaboration in education and scientific research (Tyril 2006). The Faroese Department of Foreign Affairs is working at five levels: the first deals with the fundamental question of Faroese independence, the second with trade relations, the third with international cooperation on the management of marine resources and the ocean, the fourth with international security, and the fifth with international development aid.

As to the politically sensitive question of independence, the approach agreed to by the present coalition government, is to first make full use of the *Home Rule Act* with later amendments—which means gradually assuming responsibility for any and all areas of legislation and administration pertaining to the Faroes, in accordance with existing agreements with the government of Denmark. Under this framework, the Faroes is governed largely as an independent country with national security as the principal area of exception, while it will be the responsibility of the Faroese people to decide whether and when to establish national sovereignty (*ibid*).

The political economy of the Faroes developed from settlement in the 9th century through to the present day, during which time it moved from a marginal agricultural economy to a marine-based economy fully integrated within the global capitalist market economy. During those centuries, the emergence and development of property rights to whales and the developing institution of the *grindadráp* were shaped and adapted in response to various historic changes occurring in the Faroes and overseas. What is evident is how the *grindadráp* has provided essential economic and social benefits since settlement through to the present day, for apart from providing food security, it also served to maintain strong cooperative social networks among dispersed Faroese communities.

Nevertheless, the historic record indicates that the early rules of Faroese whaling were initially inequitable, reflecting the laws and customs of a particular historic period of the political economy—one that for a long period favoured absentee landowners and ruling elites. Although the early *grindadráp* was inequitable, Faroese peasants had few options available for obtaining this valued and inexpensive food, and therefore had little choice but to accept this inequitable institution.

It was a Faroese person, Svabo, who in the late 18th century asserted that "the hunt cried out for a whaling law, which would set out to control everything having to do with the hunt and that this law should be based on the advice and experience of the oldest and wisest men" (Joensen 1976:1-2). This period marked the beginning of greater Faroese control over a uniquely Faroese common property resource institution, the *grindadráp*, so that it provided greater benefits to Faroese communities, and thereby made it a more robust institution Had this not occurred, and inequity remained, emerging economic opportunities provided by the coastal fisheries could, arguably, have drawn participants away from investing time and energy in the hunting of pilot whales which, without large-scale participation and cooperation, would have represented an enormous challenge to the continued success and viability of the *grindadráp*.

The *grindadráp* encompasses a complex inter-generational system of belief, where knowledge and values associated with long-practised resource use are passed between generations as important elements of the oral cultural traditions and customs of Faroese society. There is strong evidence of the fundamental characteristics of a subsistence economy (as described by Wolfe 1983) within Faroese society, and as is the case for all subsistence societies today and through at least the 20th century (*see* Peterson and Matsuya 1991), subsistence is intricately linked with the market economy and exist within a single society where a modern capitalist commercial fishing industry rests comfortably with traditional systems of resource use which provide substantial non-commercial networks for sharing, distributing and exchanging food and materials. The Faroese have the benefit of both subsistence and market-based economic sectors because they have worked actively to maintain the benefits derived from each (*see* Joensen 1988).

# *The* Grindadráp—*the Common Property Resource Institution*

*If we were forced to stop whaling we would lose a part of ourselves, a part of our culture and most of all lose an important part of our food. Whale meat is a big part of our food; still today we need it like we did in the past. It's important to get meat from your own area—local food, without importing it* (Ólavur Sjúrðaberg 2002, interview).

As discussed earlier, common property regimes have been shown to develop when a group of individuals is highly dependent on a resource and when the availability of the resource is uncertain or limited (Runge 1992). Such institutions are locally-devised systems of property rights that are culturally anchored in a detailed time and place (Ostrom and Schlager 1996).

*Grindadráp* is the Faroese word that describes the driving and slaughter of pilot whales. However, the *grindadráp* is far more than that, for it binds the community through a large-scale co-operative activity anchored in a cultural time and place, with the users maintaining a strong store of knowledge and understanding about the ocean, its currents, and its abundant resources. Furthermore, the *grindadráp* contributes to the maintenance of Faroese society, culture and national identity across a number of widely dispersed communities located on a number of islands. Thus the term *grindadráp* refers to the institution of Faroese pilot whaling and the various rules, regulations, and practices that guide decision-making affecting both the resource users and the resource.

The *grindadráp* has slowly developed over centuries around the use of pilot whales by the Faroese, and because of this history and the economic importance of the hunt, it is deeply rooted in and reflective of Faroese culture and the historic struggle of the people to survive and maintain their own independence and decision making over local resources. The institution of the *grindadráp* continues to provide important economic goods to all the communities across the islands, and as Wylie and Margolin note (1981:101), it is "an eminently practical business" and can only be successfully undertaken by many individuals working together and with advance knowledge of who benefits and how.

## Executive Orders Concerning Pilot Whaling

The Ministry of Fisheries is the Faroese government agency responsible for the management of whales and whaling in the Faroes. The Ministry undertakes this management task through the administration of the Executive Orders Concerning Pilot Whaling (*Kunngerð um Grind*: 2000), or what is commonly referred to in English as 'The Pilot Whaling Regulations.' The Ministry also coordinates Faroese participation in international scientific and conservation bodies that deal with the management of whale stocks (Sanderson 2006, interview).

The Executive Orders stipulate that the sheriff must send the Faroese Home Rule Government the pilot whaling accounts and a brief report of the kill and must report if there have been any breaches of the regulations. The sheriff must also report if there were orders to allow a school of whales to return to the open sea, or to stop a kill in the district (*Kunngerð um Grind*: 2000 II §25). The Executive Orders also state that for breaches of the regulations, a harsher penalty can be imposed if other laws have also been breached (*Kunngerð um Grind*: 2000 II §25). For example, in regard to the humane treatment of animals, this is covered by the *Protection of Animals Act* (1984). This stipulates that anyone who violates the *Act* will be penalised, at the very least with a fine or, for serious breaches, imprisonment.

The use of sanctions can be minimised when users consent to the rules in use, which is particularly so in regard to the Executive Orders on Whaling in the Faroes as these are seen by the users as embodying the traditional rules and customs of the *grindadráp*. In addition, the ability of resource users to modify the rules in order to adapt to changing circumstances ensures that ownership of the rules is vested in the resource users, with less reliance placed on top-down sanctions to enforce compliance.

Since the Executive Orders first came into effect, they have been amended numerous times. They were first amended in 1857 and then again in 1872 "principally only with regard to the extent of the districts taking shares in certain whale voes[1]" (Müller 1882:6). Since the inception of the Executive Orders through to the present time, they have continued to be amended to reflect social, economic and environmental influences; the most recent amendment occurred in May 2001. Updates occur to keep the management of the *grindadráp* consistent with technological developments and to refine some of the organisational aspects of the drive, slaughter, and distribution of the meat and blubber. For example, in the late 1990s an update occurred with the appointment of a committee to review the provisions for allocation and distribution of shares of the catch. This committee worked closely with district officials and community representatives responsible for whale drives in all Faroese whaling districts in order to introduce acceptable changes. The committee was seeking to resolve the issue of areas not securing enough whale meat and blubber to meet the demand. This problem arose because in some whaling districts there were many *grind* and a small population of people, often resulting in a local surplus of meat and blubber. Whereas in some other areas, such as Tórshavn with a population of more than 18,000 people, there was never enough locally-produced whale meat and blubber within that whaling district to satisfy demand. To resolve this problem, a new distribution system was introduced that would ensure there was a wider distribution of meat and blubber to those towns and districts where households do not often receive a share of the catch.

---

[1] Old Norse *vágr,* a small bay or creek (*The Oxford English Dictionary* 1995).

The new Executive Order implemented to rectify this recurring problem of supply to the capital and some other districts authorized district councils to decide the most appropriate way of compiling lists of people to whom 'the resident shares' should be distributed. Traditionally *grind* distribution lists have been based on the registry of all residents in the whaling district. However with internal migration and the growing population of the capital, the distribution system was viewed as being ineffective, as not everyone who wanted pilot whale meat and blubber was receiving a share and in some instances, those who no longer wanted or needed the meat and blubber were still receiving a share.[2]

**Figure 7.** *Tórshavn residents in line to receive their share of meat and blubber* (Bjarni Mikkelsen 2004).

In Tórshavn, the City Council (the relevant district authority) developed the new system, in consultation with the community where city residents could register for a share of the pilot whale meat and blubber when shares become available there. The new system was first used in July 2001, when 53 pilot whales became available from a total catch of 120 in the village of Sandur on Sandoy. The whales were initially flensed by the *grindamenn* on Sandoy and then packed into fish bins before being transported to Tórshavn where the meat and blubber was divided among residents on the council list with names from A to Jakobsen. There was an announcement on the radio advising people on the list with names between A and Jakobsen to come to the harbour area to receive their

[2] As with all societies, dietary preferences change over time. This is especially so with easy access to imported food products. Such a situation has occurred in some Faroese communities where younger people have a food preference for pasta, pizza and other imported foods.

share (Figs. 7 and 8). Unlike the more traditional *grindadráp,* where people are present to flense their own whales, here the meat and blubber was already cut into manageable-sized pieces and the people just went to the habour wharf with buckets and bins to take the meat and blubber home to be stored. The shares were distributed by weight, with each person getting a similar amount on an individual basis, rather than a home share.

> This new way of distribution is obviously an indication that pilot whale meat and blubber is still highly sought after by Faroese people who are living in an urban centre such as Tórshavn (Sjúrðarberg 2002, interview).

**Figure 8.** *Part of the catch from a drive in Sandoy, distributed in Tórshavn where there is a shortage of meat and blubber* (Bjarni Mikkelsen 2004).

In addition to the whales that were taken to Tórshavn on that particular occasion, a further 20 whales from the same catch on Sandoy were redistributed to the municipality of Skála where there was also a shortage of meat and blubber.

## Pilot Whaling Districts

Since the Executive Orders came into existence in 1832, the Faroes have been divided into several whaling districts. Faroese commentators (*e.g.,* Joensen 1976), say it cannot be said with certainty whether these divisions are older than the first regulations.

**Table 2. Faroese Whaling Districts and Whaling Bays.**

| **Whaling Districts** | **Whaling Bays within each District** |
|---|---|
| Norðoya Sýsla | Hvannasund, Viðvík and Klaksvík |
| Eysturoyar Sýsla: | Norðragøta and Syðragøta (counted as one), Fuglafjørður, Funningsfjørður and Norðskáli (this is also known as Sundini because you never know from which side the whales will swim when being driven) and Tjørnuvík (the same foremen officiate here) |
| Streymoyar Sýsla | Hvalvík, Tórshavn, Leynar and Vestmanna |
| Vága Sýsla | Bøur, Miðvágur and Sandavagur, (In 2006, Sandavagur was undergoing a trial and is part of Miðvágur with the same foremen) |
| Sandoyar Sýsla | Sandur and Húsavík |
| Suðuroyar Sýsla | Hvalba, Trongisvágur, Øravík, Vágur and Fámjin |

The Executive Orders (2000) divide the Faroes into six whaling districts with a total of 23 authorised whaling bays, all of which meet the requirements for suitable beaching conditions. In each district, the sheriff in charge supervises the hunt, and each authorised whaling bay has four publicly elected whaling foremen (*grindaformenn*).

Not all bays in the Faroes provide suitable topographic conditions for the whale kill. Experience over the years has shown which bays offer the best chances of a 'good' kill. The bays are chosen because of the topography of the sea-bed which must have a very gradual gradient without any steep inclines. This gradual incline ensures that pilot whales, when driven into a whaling bay, cannot sense (by echolocation or any other sense) that there is an obstacle (the shoreline) in their path, allowing them to be stranded in shallow water or on shore.

The whaling districts are based on police districts where the officer in charge of the Executive Orders is the sheriff (*sýslumaður*). The meat-sharing area is the place where meat from the different bays is first distributed if the *grind* is large enough. When a district has enough meat and blubber stored, the sheriff may still allow a *grind* to be taken so that it can be distributed to other districts where a demand exists. For example, with reference to Figure 9, *grind* from Klaksvík (green on the map) can be distributed to the area east of Klaksvík (blue on the map) and also to the west of Klaksvík, to Eysturoy (the yellow/green part of Eysturoy). Tórshavn is unique, in that it has a large population and therefore there are never enough pilot whales to distribute to other districts. While the map, in theory, illustrates the distribution of meat and blubber, in practice there are many kin networks operating, so that, *e.g.*, people living in a particular community (e.g., Leirvík) may receive a share of meat and blubber from another community (e.g., Bøur).

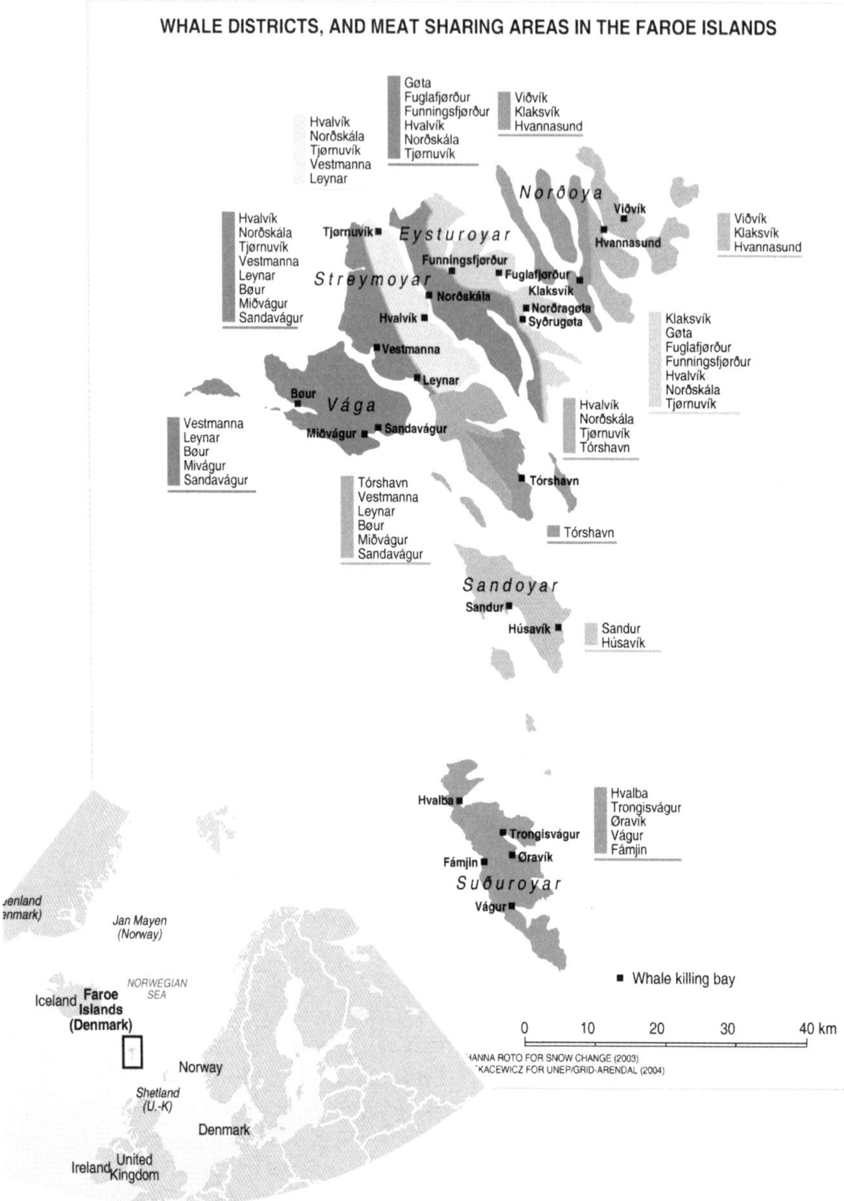


**Figure 9.** *The whaling districts (Vágar; Streymoyar; Eysturoyar; Norðoya; Sandoyar; Suðuroyar), meat and blubber distribution areas (shaded), and the whaling bays (red squares)* (UNEP/GRID-Arendal 2004).

Membership of the *grindadráp* is open to all Faroese, as all Faroese have a right to a free share of the meat and blubber. However, in practice, membership may vary, depending on what role a person plays in the drive, slaughter, assessment and division of the catch and where that person lives in relation to the whaling bay in which a *grindadráp* occurred. This variability does not pose a challenge to the stability of the institution, as this is provided for within the Executive Orders. For example, once a school of whales is sighted, property rights to the school are assigned, with the largest whale in *skinn* (or two smaller whales of equivalent *skinn* value) being allotted to the person who found the whales. Once a decision has been made between the person who found the whales and the sheriff and *grindaforemenn* concerning which bay the whales will be driven into, a wider group of users are assigned property rights to the school. However, as discussed further below, this only occurs if the school of whales is large enough to first satisfy the required allocation of shares to the boat and kill crews. A person is entitled to shares (withdrawal rights) through a number of roles they may undertake in a hunt and/or by virtue of where they live. In some hunts, a person may get a 'kill-share' as well as a 'home-share' if the size of the catch is large enough. If the *grind* is small, rights are assigned exclusively to the participants of the hunt. This is clearly stipulated in the Executive Orders and is put into practice through the share-lists drawn up in each whaling district. To be guaranteed a share, a person must directly invest time, resources (fuel and boats) and energy in the driving, beaching and slaughtering of the whales, rather than rely on a home share associated with a particular whaling bay.

Rights to the product from the *grindadráp* are transmitted between generations through either birth or migration. Each person who is either born in a particular whaling district or moves into one has a right to free access to meat and blubber. Faroese people's right to a free share of the meat and blubber of pilot whales are held for each individual's lifetime. Acknowledgement of this right is reflected in a Faroese expression which when translated to English states: "All people from the crib to old age all get one share" (Sjúrðaberg 2002, interview).

## The Sheriff and the Whaling Foremen

It is the sheriff of each whaling district who is in charge of each *grindadráp* in their whaling district. However, it is the whaling foremen (*grindaforemenn)* who play the major practical role in a *grindadráp.* From 1832 to 1909 the governor, the chief constable and the district officer were required to take charge of all whale hunts, unless weather or other unforeseen circumstances prevented them from being present, but in practice it was only the local district officer who was in charge, and prior to 1857 all the whaling foremen were appointed by the governor. After this date, they were normally elected for five year terms at village meetings (Joensen 1990).

The community nominates whaling foremen and their names are forwarded to the sheriff who then makes the final decision on their appointment. In some instances, the sheriff may disagree with the nomination and may ask for the community to reconsider their choice of nominee(s). Whaling foremen are elected and appointed according to their special knowledge and experience on various subjects related to whaling. Very often they are fishermen with an acquired knowledge of weather and current conditions in the open waters, sounds and fjords (Olsen 2006). Expert local knowledge is critical for successful driving and beaching of pilot whales.

According to the regulations, the sheriff is responsible for the administration of the pilot whale hunt and the supervision of the drive and kill, and must ensure that these activities are conducted according to the regulations which state "the sheriff together with the pilot whaling foremen must ensure that each and every person fulfils his duty" (*Kunngerð um Grind*: 2000 I §2). In practice, it is the whaling foremen who manage the operations of the hunt, on the water and at the killing bay. It is prohibited to herd pilot whales without the leadership of the sheriff and/or pilot whaling foremen. The Executive Orders state that all participants have to follow orders issued by the sheriff and/or the whaling foremen during the entire hunt (*Kunngerð um Grind*: 2000 III §8.3).

The sheriff and the whaling foremen, when overseeing a pilot whale hunt, have the Faroese flag or a special mark prescribed by the Faroese Government raised on a short pole positioned at the extreme rear of the boat (*Kunngerð um Grind*: 2000 II §7). This ensures that they are easily identified among the dozens of boats during the pilot whale hunt.

**Figure 10**. *An identification-sash worn by the* grindaforemenn (Kerins 2006).

During each pilot whale hunt there are four whaling foremen. Two undertake duties from boats in the drive itself, while two remain on shore at the whaling bay to supervise the beaching and killing. Here, they too, are required to identify themselves by possessing an identification card or a distinguishing mark (*Kunngerð um Grind*: 2000 II §7.2). This is to ensure each person participating in the slaughter of the whales knows who is in charge and clear lines of responsibility are maintained. This is important during the slaughter as the killing must be done swiftly and with up to several hundred whales being beached the situation can easily become confused. In Fuglafjøður, for example, the whaling foremen and other community people involved in the hunt, such as the assessors, have recently introduced the wearing of sashes to identify themselves (*see* Fig. 10). The *grindaforemenn* are accountable to the users of the resource. This accountability comes about through the selection process for *grindaforemenn* where the communities nominate their own people to be officials in the *grindadráp*. The sheriff is the only individual within the *grindadráp* who is paid.

While the sheriff and the *grindaforemenn* are local monitors who report back to the users, sometimes there are also international observers employed by the North Atlantic Marine Mammal Commission (NAMMCO). A veterinarian is also often present at the slaughter and hauling up of the whales. The veterinarian reports to the users, through the Pilot Whalers Association, the Faroese Government, and NAMMCO.

Another aspect of monitoring is the public nature of the *grindadráp,* for a *grindadráp* is an event with many people involved, both as participants and spectators. Spectators hold numerous vantage points from which the *grindadráp* can be observed. Such a public event, particularly one where the rules are known as well as the rules of a football match, ensures that users who may breach the rules are quickly seen and identified.

## Sighting Whales and Sending the Message

Pilot whaling can take place at any time of the year, but catches most commonly occur in July and August when the weather is more settled and there are many people on the water during the long hours of summer daylight. In June, the sun sets around 9:30pm and rises a short time later around 3:40 am. This is in sharp contrast to the situation in December, when the sun rises around 9:50 am and sets around 3:00 pm. A school of pilot whales may be spotted either from land or from a boat, the latter usually being the case within a distance from land, usually not exceeding four nautical miles.

Sending a message that a school of pilot whales has been sighted is referred to in Faroese as *grindaboð*. Once the cry of '*grindaboð*' caused whole communities to leave what they were doing and move swiftly into action, for it signified the opportunity to obtain meat and blubber to aid survival. *Grindaboð* has been described by observers of pilot whaling as "the most exciting word in Faroese" (Wylie and Margolin 1981:85). Even today, with the use of the mobile phone, the word retains its sense of excitement, as the products of the hunt—the meat and blubber—remain economically important to many households, along with other subsistence foods such as seabirds, fish, and sheep.

In the past, there were elaborate methods for relaying *grindaboð* from the person who first sighted the school, from either a boat or land, and then from village to village and across the fjords from island to island. Prior to the arrival of the telephone in 1906, the main means of communicating *grindaboð* was by runners and the use of beacons (*grindaglaða)* located at specific points on the coast and hills. Williamson (1945:120) describes such a sight near Argir, once a small village to the south of Tórshavn, where there were:

> three sites from which the warning was passed by beacon to the island of Nólsoy, four miles away across the fjord. One of these was used in the event of the whales being hunted in the fjord, another if the hunt was proceeding on the west side of Streymoy, and a third if the whales were being driven into Tangafjørður, between Streymoy and Eysturoy. A beacon was lighted by the people of Velbastaðdur to warn their neighbours at Kirkjubøur and on the opposite islands of Hestur and Koltur, as well as the inhabitants of Miðvágur. The reverse also applied, of course, and in no case was the beacon allowed to burn down until a similar signal, acknowledging receipt of the message, was seen at the receiving end. Usually the fire was made of damp hay, as this gave a column of dense smoke. In some places, the method of signalling was by laying white sheets on the grounds at the top of a prominent hill, and Sandavágur relayed the messages to Kvivík, across Vestmannasund, in this way.
>
> In districts where intervening heights rendered it impracticable to send the warning in these ways, a runner was employed to carry the news to the next village, where he was relieved by another on the discharge of his mission. Velbastaðdur and Tórshavn, and Hoyvík and Hvítanes in the South of Streymoy district were linked up in this manner in former days.

The system of signal fires, prior to telephones, influenced the distribution pattern for meat and blubber. That is why some islands are divided in two in the whale distribution map (Hoydal 2006, interview). With its introduction, the telephone slowly became the main channel for communicating *grindaboð* in the 20th century. "Indeed the [telephone] directory carries an instruction to subscribers as to the correct procedure to make this call, with the assurance of re-imbursement of any charges incurred" (Hoydal 2006, interview).

When a school of pilot whales is sighted from a boat, a piece of clothing is hung from the masthead to signal that a school has been sighted. This tradition remains in the Executive Orders, which stipulate "a piece of clothing must immediately be attached to the masthead, as is the custom" (*Kunngerð um Grind*: 2000 II §1). If clothing is not raised, then it is usually a buoy that is tied to the mast (Fig. 11).

When this signal was seen from land, beacon fires were lit, sheets laid out or turf turned at prescribed places, while runners were sent hurrying along the paths to call *grindaboð* across the narrow fjords and sounds. Boats, with either six or eight-man crews hastily rowed between islands to join the hunt (Joensen 1976). Müller (1882:6) describes the reaction to *grindaboð* in the late 19th century:

> When the message of *grind*… is noticed in a village (sometimes long before it is reported where the *grind* is, as the cry can

frequently be heard from the hills long before the messenger is advanced within a mile of the village), it is like an electric spark upon the population. The men rush immediately to their boats, only taking time to take with them their *Homlebaand* (a piece of the fin of the whale dried for a sling for the oar). Their women run to the shore, carrying necessary articles of dress and food for them. The children cry '*Grindaboð*' in the mouth of each other, and a stranger might believe that the whole population had become insane, as it is not sorrow or fear for any calamity that is shown, but joy that is shining on every face.

**Figure 11.** *The buoy tied to the masthead signifies the finder of the* grind. (Kerins 2006).

Today, because of technology the call of *grindaboð* is not heard as much as in the past. Prior to the animal rights protests, the sighting and whereabouts of a *grind* was reported on Faroese radio. Because of possible interference in the hunt this was stopped during the 1980s, with the radio reporting only at the conclusion of a successful *grindadráp*. The urgency of getting the message out to as many people as possible in the whaling districts is an important task. A high degree of social cohesion underpins each *grindadráp*. It is important to have enough boats in place as soon as possible to ensure that the whales are contained and do not escape. Employers usually make allowances for members of their staff to take time off during drives so that they may participate.

When the message of the sighting is relayed from the finder to the sheriff, he, under the Executive Orders, becomes responsible for alerting the pilot whaling foremen in the district where the whales have been sighted. The pilot whaling foremen then quickly spreads the message throughout their district. The sheriff is also responsible for advising if there is a government prohibition on pilot whale hunting in place within that district.[3] This is undertaken when the whaling foremen, in consultation with the community of the whaling district, decide that any further take of whales would create over-supply and cause wastage. For example, in the period from May 1986 to December 1988, several whaling bays were closed on 11 occasions to avoid an over supply of meat and blubber (Department of Fisheries, undated:15). As discussed above, some whaling districts, for various reasons, have a higher occurrence of *grind* and therefore have an abundant supply of pilot whale meat and blubber. Whaling districts which have lower rates of *grind* often run low on supplies.

> The areas that often get *grind* are well known. They are the northern islands, in Klaksvík, in the western area in Vagar, there are three whaling bays on Vagar, at Bøur, Midvágur and Sandavágur and a small population about 4,000 people, so this area often has a surplus. When *grind* occur close together the local sheriff will arrange for the *grind* to be distributed elsewhere (Sjúrðaberg 2002, interview).

## The Drive

The driving of the pilot whales is called *grindarakstur* in Faroese and the hunters are called *rakstrarmenn* (pursuit men) or *grindamenn* (whale men). Each boat in the hunt must also designate a foreman. This is usually the owner of the boat, but a substitute for the owner can also be considered the foreman. Both the sheriff and pilot whaling foremen are required under the regulations to know who this person is (*Kunngerð um Grind*: 2000 I § 3). The foreman of the boat, in turn, has the responsibility of ensuring that all crewmembers fulfil their duty (*Kunngerð um Grind*: 2000 I § 3.2). Each boat is also required to carry its harbour registration mark (*Kunngerð um Grind*: 2000 I § 3.3). The pilot whaling foreman of the boat is required, prior to each hunt, to ensure that his boat has all the equipment on board. Under the pilot whaling regulations this includes: stones fastened to lines; a gaff and/or a ball pointed hook attached to a rope; a whetstone; and a pilot whaling knife (*Kunngerð um Grind*: 2000 II §6–6.2)[4].

Pilot whale hunts only take place when a school of whales is sighted close to the coast. A drive begins usually no more than two to four nautical miles from

---

[3] The Government can impose a temporary ban on whaling if it is considered that the households in the whaling district have enough meat and blubber for their needs. The ban, however, is not enforced until other whaling districts consider their households' supplies, for others may request whales continue to be killed and the shares distributed in their district to replenish their supplies (*Kunngerð um Grind*: 2000 I § 1.2 - 3).

[4] It is prohibited to use any other equipment than those outlined in §6.2 of the Executive Orders. However, with special governmental permission, it is permissible to use equipment on an experimental basis with the sole purpose of improving the killing method (*Kunngerð um Grind*: 2000 II §6.3).

the coast, and only when the sea and weather conditions make it possible to drive the whales safely to shore. If the whales are too far offshore, they have more opportunity to outflank the *grindamenn* by splitting into smaller schools and leave the *grindamenn* hoping that they are pursuing the larger group. The secret to a successful hunt, say the *grindamenn*, is to guide the whales rather than force them in a particular direction. When I asked a pilot whaler (*grindamaður*), "why did the Faroese still rely on opportunistic sightings of *grind* when motor-boats now enable people to go looking further afield for whales?" he replied:

> It is too difficult to drive them to the bays when they are too far away. Normally they are not more than 2 to 4 nautical miles from the whaling bays. You don't start driving them much further than this. This is tradition and it's hard to change tradition. We know that if you drive the whales for too long they get stubborn, it's like they work out what's happening and the drive may not be successful (Hermanson 2002, interview).

During the two-year period of an international research project[5] carried out on pilot whales, there were 47 notifications of *grind* recorded, of which 43 resulted in beachings. One third of the schools were first observed from land and the distance from shore ranged from 0.1 to 3.3 nautical miles, with an average of 0.9 nautical miles. About 40 percent were observed between 9:00 am and 12:00 noon, with 80 percent sighted between 6:00 am and 3:00 pm (Bloch *et al.* 1990).

The driving of pilot whales begins as soon as a sufficient number of boats has reached the whales and loosely form a semicircle behind the school. The school must be herded into an approved whaling bay in accordance with the Executive Orders (*Kunngerð um Grind*: 2000 III §9). The sheriff and the *grindaformenn,* after consultation with the boat that found the *grind,* decide the direction the drive will take and the whaling bay that the whales are to be driven

[5] In 1986 a large international research project on long-finned pilot whales in Faroese waters was undertaken. It was initiated and partly financed by the Faroese Government, under the auspices of the United Nations Environmental Programme (UNEP) and the International Whaling Commission (IWC). Scientists from the Faroes and a number of research institutes in several different countries, including Denmark, France, Spain, and the U.K participated, along with much help from Faroes whalers. Intensive sampling of a total of 3,488 pilot whales from 40 different schools was carried out between July 1986 and July 1988. The unique nature of the catch in the Faroes where whole schools are taken, allowed the scientists the opportunity to gather large amounts of valuable biological data which had not been possible to obtain before, without any specific selection according to sex, age, size or other factors. The fact that whaling occurs throughout the year in the Faroes also provided the opportunity for scientists to study the social structure and mating habits of the pilot whale. The research focussed on stock identity, reproduction, social structure, feeding, age-related parameters and the presence of parasites and pollutants in the whales (Department of Fisheries, undated: 6-9).

toward from among the various whale bays. The direction of the current plays a large part in the direction a *grind* will be driven (*Kunngerð um Grind*: 2000 III §8). According to the Executive Orders, the most convenient whaling bay has to be chosen according to the size of the school, the circumstances and the tidal current conditions (*Kunngerð um Grind*: 2000 III §8). Instructions from the whaling foremen are conveyed by means of on-board radios or sometimes by megaphone. When enough boats have arrived to join the finder's boat, the drive begins under the command of the two *grindamenn* out on the water. The whales are initially driven steadily and without pressure or the noise of shouting. The pilot whalers always drive the whales with the current; *grindamenn* often say it is a matter of local knowledge that you cannot drive whales against the current and you must let the whales set their own pace or the drive will be unsuccessful (author's field notes, August 2001).

These same actions have been described in the late 19th century (Müller 1882). Other factors can also come into play, and in one of the *grindadráp* in which I participated, consideration was also given to the quality of road access to the first whaling bay chosen, as the amount of traffic expected the following morning during the division and distribution of the catch would exceed the narrow road's capacity. The *grindaformenn* were also of the view, because of the late time of day, that the school of pilot whales may have had to be held offshore from the proposed whaling bay overnight due to the conditions of the current in the channel, and the proximity of the whaling bay at Tjørnuvík to the entrance of the sound which could allow the whales to escape to the open sea. Because of these considerations, the pilot whaling foremen decided to drive the whales further into the sound and beach them at the whaling bay at Norðskáli. It is for such reasons that some of the routes used to drive whales may not appear very direct when drawn on a map.

There are numerous factors that affect how a *grindadráp* proceeds, such as the time of day, the location where the school was sighted, the direction and time of the tides and wind. The manner in which some schools behave also plays a part in the length of time taken to complete a drive. Some schools of pilot whales surface and dive in an almost systematic pattern, moving quickly in the direction they are being driven; others move more slowly, taking longer to surface and do not behave in such a regular fashion. During the 1984–86 international research project, the duration of 43 drives ranged from 15 minutes to 10 hours 45 minutes, with the average duration of a drive being 2 hours 22 minutes (Bloch *et al.* 1990). During the drive, the smallest boats are in the front and the larger boats are at the back, usually in two or three rows, forming a barrier to prevent the whales from escaping (*Kunngerð um Grind*: 2000 III §10.3). Schools can sometimes be held completely still by the boats at a distance from the shore, waiting for suitable conditions to arrive to drive the whales into a whaling bay. This practice is old and was described by Müller:

> When the tide is unfavourable for driving the *grind* towards the whale voe, where it is intended to capture it, the herd must be "laid by," that is kept in a bay where there is little or no tide till the favourable tide sets in. A *grind* may be kept for other reasons (Müller 1882:6).

Once a school of pilot whales has come into an authorised whale bay, the sheriff and/or the pilot whaling foremen who belong to the whaling bay have to decide when and where the whales will be driven ashore, and how the kill is to be conducted (*Kunngerð um Grind*: 2000 III §10). On the pilot whaling foremen's signal, stones, both loose stones (*leysakast*) or stones or weights attached to lines (*fastakast*) are thrown into the water behind the whales, driving them toward the beach where they become stranded. According to the Executive Orders, any group of whales that cannot be beached in this manner must be driven out to sea again (*Kunngerð um Grind*: 2000 III §11.2).

During the two-year period of the international research project, 43 *grind* were taken, with three of these being kept overnight near the shore before being driven ashore at dawn. Two of these resulted in a difficult beaching and the third was released by order of the sheriff, after several unsuccessful attempts at beaching the school (Bloch *et al.* 1990:38).

## Slaughtering the Pilot Whales

Prior to the whales being beached, the sheriff and/or pilot whaling foremen are required by the Executive Orders to ensure a sufficient number of people are ready on shore to kill the whales. It must also be ascertained that the area around the whaling bay is free of unauthorised people, permitting the pilot whalers to do their work unimpeded (*Kunngerð um Grind*: 2000 III §10.2). "The efficiency of the kill is largely dependent on the overall organisation of the drive, from the time it is sighted and the whales are driven toward the shore, as well as the experience and skill of those on shore who carry out the work of killing the whales when they have been beached" (*Grindaboð* 1996:3). When the whales have been beached or are very close to the shore, the sheriff and/or pilot whaling foreman give orders to kill the whales. A whale which has not beached may be hooked and hauled ashore. However, it is not permitted to hook a whale unless the sheriff and/or pilot whaling foremen have given special permission (*Kunngerð um Grind*: 2000 III §10.4).

Schools of pilot whales can sometimes number many hundreds of animals. If the school is larger than the available space permits in the whaling bay, the sheriff can stop the kill. Arrangements then have to be made to drive the remaining whales out to sea. In the case of several unsuccessful attempts to beach the whales at the appointed whaling bay, and if the use of hooks or gaffs is not possible, the sheriff, together with the whaling foremen, are required to order that the whales be driven back out to sea (*Kunngerð um Grind*: 2000 III §10.5).

The Executive Orders prescribe the killing method (*Kunngerð um Grind*: 2000 III §11.2): the whales must be killed by making a dorsal cut, from one side of the head to the other, severing the spinal cord. A second cut is then made to sever the smaller minor veins to bleed the animal (Fig. 12). These cuts can be made within a matter of seconds. "A dorsal cut is made in the neck about a hands breadth behind the blow hole, deep enough to sever the spinal cord and surrounding [blood] vessels leading to the brain, at which point the animal is paralysed and loses consciousness within 5–10 seconds" (*Grindaboð* 1996:3).

**Figure 12**. *The prescribed killing method: a cut, one hand-breadth behind the blow hole.* (Bjarni Mikkelsen 2004).

Much Faroese and NAMMCO effort has gone into studying the slaughter of the whales. To some degree, this has been influenced by the protest action. Jústines Olsen, a senior Faroese veterinarian, who has been conducting research on humane killing methods for pilot whales, describes various methods that have been studied, reporting that:

> We have considered alternative methods, in particular shooting. But in the pilot whale hunt there are many people working together quickly in uncontrolled conditions. For practical and safety reasons, the use of firearms is not an option. Given the circumstances of the pilot whale hunt in the Faroes, the centuries old traditional method of severing the spinal cord, which also disrupts the major blood supply to the brain, is the quickest, safest and most effective way to get the job done (*op. cit*).

The driving of whales ashore and swiftly killing them was made more difficult with the construction of harbours, seawalls or quays, and more recently the development of salmon farms in the sounds and fjords. These often obstructed the beaching of the whales, making it necessary to kill the whales in deeper water (Hoydal 2006, interview). Following a review of the condition of each of the whaling bays used in the beaching and slaughter of whales, several changes were made. In some places, boat sheds and their concrete slipways were removed to ensure there were no obstacles to entangle and distress the whales.

The two-year international research project identified a link between the ratio of men to whales and the time to slaughter. No specific correlation between the killing time and the number of whales was found. The shortest kill took place on the sandy beach of Leynar and required eight minutes to kill 136 whales. The longest kill was at Vestmanna where 118 whales were killed in two hours and 30 minutes. The average killing time for 40 pilot whales was 27.7 minutes (Bloch *et al.* 1990:40).

In an ideal situation, most of the whales will be stranded far enough up on the shore that it is unnecessary to secure them. However, those remaining in the shallows must be secured and hauled closer to shore. Traditionally this was undertaken by driving a steel hook, or gaff, with a rope attached to it into the back of the whale. Men then hauled on the rope and pulled the whale in to shallower water or ashore. Prior to the invention of the hook, the whales were secured simply by holding the fingers in the blowhole.

During 1995-1996, the Faroese Government again monitored the performance of the hunt in a systematic way by timing the specific phases of the kill in order to assess where time-to-death could be minimised under the prevailing conditions in which the work is carried out. The time was measured from the moment an animal was first wounded by the gaff or knife, to the loss of consciousness. It was found that there was a significant difference in the time it takes depending on whether or not the whale needed to be secured with the gaff before killing. This killing method was then examined to determine if improvements could be made (*Grindaboð* 1996:3). The result of this examination was the development of the new blunt hook (*blástrarongul*) which did not wound the animal prior to killing it. Olsen concluded that "The average times-to-death in the pilot whale hunts are as good as or better than other forms of hunting for which there is comparative data. Since we began registering data, the maximum time-to-death using the gaff was 5.5 minutes, and 2 minutes using the new blunt hook" (*Grindaboð* 1996:3). Olsen, comparing the killing of pilot whales with that of other animals, stated:

> There is very little common ground for comparing the slaughter of domestic livestock with the hunting of wildlife. Technical slaughtering systems are constructed for domestic animals, and large investments are made in scientific research aimed at making the slaughter as humane as possible. The aim is of course that animals should die 'instantaneously,' but even with the most advanced equipment, this ideal is often compromised in practice by other factors. Domestic animals are totally dependent on the conditions imposed on them by the farmer, from the moment they are born until they are slaughtered. Demands from livestock productivity and economic returns can easily compromise general standards of animal welfare, often resulting in increased suffering and stress due to cramped living conditions, genetic programming for higher yields, and long-range transport to slaughter.
>
> One of the drawbacks in comparing our methods with others is that there is very little comparable data except from other forms of whaling, such as Norwegian minke whaling, even though most

> countries today have some form of wildlife hunting. If we had similar data from, for example, kangaroo hunting in Australia, or deer hunting in the U.K., then we would have a far better basis for comparative discussion of international standards (*op. cit*).

It is also difficult to compare the pilot whale hunt, in which a whole school of whales is killed at once, with other forms of hunting in which a single hunter takes aim from a distance and shoots the prey. With the simple technology of the knife used to kill pilot whales, men come in close physical contact with animals, and no wounded animals escape or are left to die slowly (*op. cit*). This subsistence mode of production ensures that Faroese people are acutely aware of what is involved in relation to the slaughter of animals for food. In this activity, evidence for the persistence of a domestic society, as described by Bulliet (2005), is strong.

## Hauling-up and Flensing the Whales

As soon as the whales are dead, the sheriff and/or the whaling foremen immediately arrange for people to keep watch, both on the sea and land and give the watchmen whatever instructions are necessary. Their duties include keeping the whales intact—except for opening the belly to cool the meat so it does not spoil—and ensuring that no unauthorised person has access to the whales. They also make sure that any whales that may have sunk can be hauled ashore should they float to the surface (*Kunngerð um Grind*: 2000 IV §12). Sometimes, some of the dead whales sink. It is prohibited under the Executive Orders for anyone to bring them to the surface without permission from the sheriff or the pilot whaling foreman. All other boats are obliged to stay away when this occurs (*Kunngerð um Grind*: 2000 IV §13.5).

Once the kill is over, the Executive Orders stipulate all whalers and boats are obliged to assist in hauling the whales to the nearest wharf, or if a wharf is not available, to haul the whales up above the high-tide mark on the shore in such a way that the heads are free from the water at high tide (*Kunngerð um Grind*: 2000 IV §13.2-3). A wharf is usually the preferred place as the whales can be laid out on a concrete surface, away from sand and gravel, so the meat and blubber does not get contaminated and the area can be hosed down with cold water. A wharf also makes access easier when the crowds arrive to flense the whales after they have been assessed.

To haul the dead whales to the wharf, the *grindamenn* usually secure the whales by their tails with rope and then drag them through the water to the wharf where they are lifted out of the water by a crane. On occasion, when there are larger fishing boats with cranes present, several whales may be hauled up on to the deck and transported to the wharf. This work is always done as quickly as possible. A senior veterinarian, under the direction of the Minister of Fisheries, is often present at a whaling bay to observe the kill and later, at the wharf, to inspect the whales as they are pulled from the water. The veterinarian checks to see if the whales have been injured in the hunt by the boats' propellers and also looks for diseases and assesses the physical condition of the animals. The veterinarian observes the slaughter, examining the time of the kill as well as the condition of the whales as they are hauled from the water, where knife marks reveal the skill or otherwise of the hunters. After monitoring the slaughter the veterinarian reports to the *Grindamannafelagið* (Faroese Pilot Whalers Association) and the Faroese Government. The veterinarian often attends the regional and national meetings of the *Grindamannafelagið* where he reports his findings. Any issues raised at this local level concerning the slaughter allow the whalers to take corrective measures rather than waiting for them to be imposed from above.

Once this work has been completed, the pilot whaling foremen order all the *grindamenn* ashore, where the names of all the participants in the drive and kill are registered. The registration has to be conducted no later than one hour after all the hauling work is completed (*Kunngerð um Grind*: 2000 IV §13.4). If there is any disagreement as to who is to be registered as a participant in the pilot whale drive, the sheriff will make a decision after consulting the pilot whaling foremen (*Kunngerð um Grind*: 2000 IV §14).

A *grindadráp* is a large social occasion, and in some whaling bays where *grindadráp* are not so common, many people will gather on the shore to view the catch and talk with neighbours and friends from other villages in the district. As the majority of *grindadráp* occur during the summer and autumn months when many expatriate Faroese are home visiting family and friends, the occasion can often be described as festive. Children are often present at the end of a *grindadráp*, and are always full of excitement about seeing a whale close-up and being able to touch and inspect it. They always have many questions about the animals, their teeth, the marks they may carry, their long flippers or how they were killed. It is on occasions such as these that the transfer of ecological knowledge between generations takes place.

## Assessment

In the past, after the kill, one of the first actions of the sheriff was to designate a *matarhvalur*, a whale to be proportioned out among households of the village where the slaughter took place so that everyone involved in the hunt could have a meal of fresh meat and blubber to warm themselves. Prior to roads and motor boats, the hunt could take many hours with the men becoming exhausted and wet from rowing and killing the whales. However, following the building of roads and the introduction of cars, many people now drive home after a hunt to get changed, so the *matarhvalur* is no longer assigned. Also in the past, as soon as the whales were opened to cool, people were also free to cut the kidneys and liver—both then considered delicacies—from the whales, as these quickly spoil. However today, because of concern about contaminant loading, these organs are no longer eaten.

Once the whales have been hauled up on shore, the sheriff appoints two or more independent people to undertake the assessment of each of the whales. The assessment of the whales must be conducted as quickly as possible following agreement between the sheriff and the people chosen to do the estimation (*Kunngerð um Grind*: 2000 IV §17). The assessment is undertaken according to the traditional and authorised measurement of *skinn*, which is an old Nordic measurement value that derives from agricultural practices

(*Kunngerð um Grind*: 2000 IV §16). Whales in the Faroes have been assessed according to the *skinn* value since 1584 (Bloch and Zachariassen 1989). This extensive record, shortened only by a 70-year period without data, allows comparisons between past and current observations and changing trends. Analysis of these statistics in the 1980s revealed long-term cycles in catches of approximately 120 years (Joensen and Zachariassen 1982), with annual catches ranging in number between 0 and 4,325. An average of 6.9 schools and 988 whales have been landed annually in the period 1709 to 1989; in the 1980s the annual mean catch was 2,064 whales (Bloch *et al.* 1990).

**Figure 13.** *The assessor and helpers using the traditional assessing rod.* (Kerins 2006).

The assessors estimate the *skinn* value of each whale with the traditional assessment rod, measuring the body of the whale from the eye to the anus. (Fig. 13). The assessment rod is marked according to a calibrated logarithmic scale (Fig. 23) representing the value in *skinn*. One *skinn* is equivalent to 38 kilos of meat and 34 kilos of blubber. "The definitive value of each whale is assessed according to this measurement, but the quality of the whale is also considered, *i.e.*, stoutness, wounds, pregnancy, lactation" (*ibid*: 41). In addition to the assessment rod measurement, the assessors also take the metric length of the whale; this metric measurement is used for scientific purposes and plays no part in the division of the catch (*Kunngerð um Grind*: 2000 IV §17.2).

Once the assessors have assessed each whale they cut the *skinn* value in Roman numerals on the whales' pectoral fins (Fig. 14), and the consecutive number of the whale is also cut, in Arabic numerals, onto the whales' heads (Fig. 12). The assessment is undertaken as quickly as possible, and by using the assessment rod ensures that the process is much faster than weighing each whale which, for large *grind,* would be very time-consuming. Once the assessment is complete, the sheriff announces what time the division will be announced. In some districts, the assessment is done manually and in others computer spreadsheets are used.

**Figure 14**. *The* skinn *value of the whale, in Roman numerals, cut into the pectoral fin.* (Kerins 2006).

## The Division of a *Grind*

In each district, in readiness for *grind,* the sheriff draws up pilot-whaling rolls (*grindalisti*) which tabulate the population in that whaling district, and how distribution will be undertaken between boat and kill shares, or by any other unit of distribution. The sheriff can, in consultation with every local municipality, decide how the pilot whaling rolls are arranged: e.g., if the municipality's population register is to be used or if any other distribution system is to be used. The final list is presented at a village meeting (*grannastevna*), and at that time people can suggest changes (Joensen 1976). The pilot whaling roll must be assessable for public viewing and is usually kept at the sheriff's office (*Kunngerð um Grind*: 2000 I §2.2).

The division of a *grind* today follows the custom of the past and consists of first deducting a row of fixed costs (discussed further below) and, if the *grind* is over a certain size, dividing the remainder among the residents of the whaling district. When a *grind* is small (in relation to the size of the community), e.g., less than 75 animals, the catch may be shared only among the actual participants of the hunt, *viz.* those who drive and killed the whales. If the *grind* is larger, the catch is also shared in equal 'home parts,' meaning that each member of a household in the whaling district, regardless of age, sex and actual presence on the islands at that time, automatically receives a share, which is brought to the village. According to the Executive Orders, only persons older than 14 years are entitled to a participating (hunting) share (Bloch *et al.* 1990:42).

The Executive Orders clearly set out the division of the catch, detailing who gets what amount of meat and blubber, according to the role each played in the hunt. Thus every participant knows in advance what the division method will be. This system reduces conflict and ensures cooperation, which is one of the most important elements in a pilot whale drive. The division rules also have some flexibility to allow the sheriff additional decision-making powers under certain conditions, such as if the drive has been long and difficult. Here, participants are provided with a greater share to acknowledge their longer hours of work. Provided the sheriff, pilot whaling foremen and whalers agree, the division may be carried out according to local tradition. The same applies to damage refunds and cleaning-up. In each case, the government has to be notified (*Kunngerð um Grind*: 2000 IV §18.2).

Once all the whales have been assessed and numbered, they are divided and distributed by the sheriff according to the Executive Orders (*Kunngerð um Grind*: 2000 IV §18) as outlined in Table 3 below.

**Table 3. Share Divisions (*Kunngerð um Grind*: 2000 IV § 18).**

| | |
|---|---|
| Finder's whale | A boat that sights a school of pilot whales and follows the school until the message has been passed on and other boats have arrived at the spot will have the finder's whale. When a school is first sighted from land or from a boat that does not stay by the school until the other boats have arrived, but the message has been passed on, the finder's whale is divided between those who sent the message and the boat which arrived first at the school. The same applies when the message is passed from one village to another, and a boat from the later village is the first to arrive at the school. Those who are entitled to the finder's whale are free to choose either the largest whale or smaller whales totalling the same total measure as the largest whale. If several people sight a school simultaneously they are entitled to share the finder's whale. |
| Watch Keeping | During the day: ½ *skinn* to each boat which is used, and ½ *skinn* to each person; at night: 1 *skinn* each. |
| Estimators | 2 *skinns* for every 100 whales, or otherwise as the sheriff sees fit. |
| Sheriff | 2 percent of the total number of *skinns*. |
| Foremen | 1 percent of the total number of *skinns* (shared among the 4 foremen 0.25% each) |
| Damage surveyors | At least 1 *skinn* each as remuneration, or as the sheriff may determine. |

| | |
|---|---|
| Division assistants | The sheriff pays remuneration according to labour and responsibility. |
| Kill share | Every person who has actually taken part in the drive and/or the kill in a boat or onshore has the right to a kill share. The kill share is the compensation, which the participants receive for their work, such as the driving or killing, fishing-up, and towing.<br>A kill share is given to everybody on a boat. The Executive Orders stipulate the maximum share per boat:<br>Eight man boats and larger boats: maximum 8 shares.<br>Six man boats: 6 shares.<br>Four man boat: 5 shares.<br>Boats smaller than above mentioned: maximum 3 shares.<br>Boats and their crews and people from outside the district are entitled to shares, provided they have participated in the drive or the kill. In order to receive a kill share a person must be at least 14 years old. A whaler is only entitled to one share even if he has driven with one boat and then participated in the killing from the shore. After consultation with the pilot whaling foreman, the sheriff can allot an extra share to the driving boats, when the drive has been particularly long and difficult.<br>In cases where a school of pilot whales has been held close to the shore waiting for the right conditions to beach the whales, the boats which did not take part in the drive, but only the kill, may receive a kill share, if the sheriff or pilot whaling foreman has asked them to participate. The sheriff determines the size of the share. |
| District Share | These whales shall be set aside for municipalities or other organisations, which shall themselves arrange distribution in the villages. The district shares shall be divided according to the pilot whale roll. If difficulties arise from the division in a district, the sheriff has the authority to divide and distribute the meat to a certain municipality or certain municipalities. This also includes municipalities outside his district. |
| Surplus whale | Sometimes it is necessary to compensate for sick and spoiled whales. If not all are distributed, the sheriff must sell the surplus in the best possible way to cover the costs for damage, cleaning and insurance. |
| When the whales are not divided into district shares. | If the district shares are not at least 1 *skinn* per boat (50 people) after deductions have been made, the remaining will be divided among the whalers. |

In most districts it is also customary to allocate a share to the local hospitals and other institutions, such as nursing homes and day-care centres. In a few of the bigger villages and towns, a small amount taken from the private shares is sold to the local food stores. The maximum retail price of whale meat and blubber was, until recently, regulated by governmental order and was about half the price of other meats such as beef or lamb; however, this price-setting practice ceased in 2003.

The cost of any damage that may have occurred to the boats or equipment, as well as the clearing and removal of remains from the beach or the foreshore where the whales were killed, and the cleaning of the wharf or any other site where the whales were laid out prior to the division being carried out, are also paid proportionately, according to the number of *skinn* in the kill and district shares of the catch. It is the responsibility of the local municipality where the catch takes place for clearing all remains within 24 hours of the division of the catch.

If whales are sold[6] to local food stores, the money so obtained can be used for material damages and cleaning up of the remains. Money may also be used for enterprises promoting whaling. If the sale of these whales gives a surplus, the money goes into a fund to support similar activities at later kills in the same district. If there are no whales to be sold then the removal of pilot whale remains in individual villages is the responsibility of the relevant municipality, which must bear the cost (*Kunngerð um Grind*: 2000 IV §19). The Executive Orders also stipulate that municipalities in the whaling district are obliged to jointly pay a collective accident insurance against personal injury that may occur during whaling. Payment is made according to a statement prepared by the sheriff (*Kunngerð um Grind*: 2000 IV §19.2). If there is an accident, the sheriff has the authority to make arrangements with the municipality in the whaling district for the immediate payment of compensation for damage incurred during the hunt, or for lost equipment (*Kunngerð um Grind*: 2000 IV §19), including:

- damage to boats, including damages to propeller or axle, oars and tillers;
- broken and lost equipment;
- lost and broken spectacles and dentures;
- personal injury.

## Flensing

Once the deduction of fixed costs has been made as outlined above, the distribution of the shares occurs. The free division of the meat and blubber of pilot whales after a successful hunt is known as the *grindabýti*. Usually, crowds begin to assemble just prior to the sheriff's prescribed time for an announcement of the division for distribution among the whaling district (Fig. 15).

---

[6] These whales include the surplus whale that has been set aside and then not used, or whales which have sunk but then resurfaced within two hours of the division tickets being issued.

**Figure 15.** *People gathering to hear the sheriff announce the division of the day's catch.* (Kerins 2006).

The majority of people who attend the division are men, often accompanied by children, who bring with them colourful buckets, fishing bins, wheelbarrows, bags of salt, knives and meat hooks, which are all employed in flensing and later storing the whale meat and blubber. The people stand about in small groups talking, and when the sheriff appears, the crowd gathers around him to hear how many whales have been killed and how many *skinn* at which the school was valued. The sheriff then calls out the Arabic number of the whale, starting at one, and then calls the names of people who will receive a share from that particular whale. One of the named people collects the ticket while the others who have been assigned to that particular whale gather around it to begin the flensing (Fig. 16). The Executive Orders require that the shares must be removed within an hour after the tickets have been issued. Any person allotted a share, regardless of his function, must cut it up himself; "One single whale is usually shared between several men, who first cut up the whole whale together and then go on to divide it into equal portions of quantity and quality" (Bloch *et al.* 1990:42).

The business of flensing the whales is another highly social occasion. Compared to the drive and slaughter, flensing is a very calm affair. There is no mechanised processing of the whales, all the flensing is undertaken by hand. The blubber and meat that is to be consumed is taken from the killing cut in the neck to about the level of the anus. Although there is meat below the anus on the thin part of the carcass immediately anterior to the tail fluke, this is not used as it is considered too tough to eat. There is a Faroese saying which reinforces this practice: *"Frá høvdinum skalt tú hvalin skera"* "From the head shall the whale be butchered," (Olsen 2006:42). Each whale is flensed quickly and always in the same manner: first the dorsal fin and flippers are removed and discarded, as these parts are no longer used.[7] The whale is then laid on its side and a long cut is made along the back of the whale, followed by about four equal transverse strips about 30 cm wide, from the middle of the dorsal to the medial ventral line, after which the blubber is virtually peeled off the carcass. At the abdomen end of the transverse strip a small slit is made in the skin where someone inserts either a meat hook or their fingers and then pulls the heavy flap of blubber up toward the dorsal fin while another person cuts underneath to free the blubber from the dark red meat.

**Figure 16**. *Each person receiving a share is expected to assist in flensing their assigned whale*. (Bjarni Mikkelsen 2004).

The person who cuts the blubber away also tries to ensure that some meat remains on the blubber as this gives the blubber a slightly pink colour when the blubber is cured in salt. The blubber is laid on the ground, skin side down, to form a mattress where the meat can be laid so as not to touch the ground and become contaminated. Once the layer of blubber has been removed, the meat is cut off in large blocks and then tossed onto the blubber. Someone, usually an older man, keeps the pile in order and cuts large blocks of meat into smaller, more manageable sizes that will be easier to share. When one side of the whale

[7] Today only the meat and blubber, and sometimes the ribs, are taken from the whales. In the past many other parts were used such as the kidneys and liver for food, while the stomachs were preserved for floats and inedible blubber was processed into oil for lighting. Parts of the skin were used for rope, the oesophagus was processed into leather for shoes and the penis dried and cut into strips which were used for thread for sewing skin shoes (Olsen 2006).

has been flensed, it is rolled over and the process starts again. When the process has been completed, approximately 54 percent of the total weight of the animal will have been processed in order to be utilised for food (Bloch and Zachariassen 1989).

**Figure 17**. *After the whale is flensed, the meat is placed on a blubber mattress* (Bjarni Mikkelsen 2004).

When flensing has been completed, there is a large pile of red meat resting on top of a mattress of thick blubber (Fig. 17). The next stage is the division between the men assigned to each particular whale, or as is the case with the new distribution system in Tórshavn, people line up and receive their share (*see* Figs. 7 and 8). The meat and blubber is stored in several different ways: meat is frozen or hung outside to dry in the wind, and the blubber is either dry-salted or wet-salted until it is ready to eat. Dried meat and blubber can be stored without refrigeration for extended periods of time.

## The Technology—Past and Present

The Faroese have always employed a small set of relatively simple tools when pilot whaling. While some of these tools have remained unchanged throughout the centuries, others have been refined or otherwise adapted, or even cast aside, in favour of new tools or techniques which are more efficient or humane. The majority of these technological adaptations have occurred over the last 50 years and have focused on animal welfare issues. The technology required in the pilot whale hunt is relatively cheap and accessible to all people on the islands. Even if a person does not have a boat, they can participate onshore, either killing the whales using the knife, or securing them using the hook and hauling them ashore. The simplicity of the technology should not be taken as evidence of an archaic institution, but rather, as a reflection of the most cost-effective and humane tools for the job of securing high value meat and blubber. In addition, the simplicity and relatively low cost of most of the tools employed in the hunt acts as an incentive for each whaling district to be fully prepared for the sudden arrival of pilot whales.

**Boats**

In the Faroes, no-one is a whaler by profession, nor is there any special 'whaling boat'; traditional Faroese coastal fishing boats (Fig. 18) have always been used in whale drives. In the past, a boat was of no use in a whale drive unless a crew of men was available to row it. However, today any type of boat can be used in a drive, from small traditional six- and eight-man boats to small commercial fishing vessels; the use of motorised boats is the most significant change in the pilot whale hunt. In the past, the rowing of the boats—first from the villages to where the whales had been sighted, a distance of usually four nautical miles—and manoeuvring the boat into position behind the whales and driving them to the most appropriate whaling bay required a huge amount of human effort.

**Figure 18.** *Typical Faroese boats used in whale drives* (Kerins 2002).

The introduction of the motorboat did not alter the drive nor the operation of the hunt, but it did significantly reduce the human effort needed in earlier times. Neither did the introduction of the motorboat increase the number of whales taken, but it did increase the number of boats and the number of people participating in the drive, which affected the division of the catch. Motorized boats provided the opportunity to greatly expand the hunt, from an opportunistic to an active hunt, where men could go to sea in search for whales and then drive them back to shore. However, this did not occur, because the Faroese whalers believe that pilot whales become stubborn if driven too far, making the drive difficult or unproductive for whales driven for longer distances and periods of time in open waters have more opportunity to escape by splitting up into smaller schools. What is evident here is that operational controls over harvesting are based upon social and cultural norms and expert knowledge about the resource, rather than restrictions placed on available technologies employed in harvesting.

### The *Kast*

The *kast* (*leysakast* and *fastakast*) are stones, which are cast into the sea to keep the whales moving in a forward direction (Fig. 19). *Leysakast* are ordinary small stones taken aboard the boats in cardboard boxes and thrown overboard behind the school of whales, and not at them as some observers believed. Müller (1882:7) described the use of the stones in the drive in the late 19th century, and the thinking at that the time regarding their use:

> The splash in the water, and the white air-bubbles occasioned by the stone when going down, the *grind* shuns to such an extent that when a regular and constant casting or throwing is kept up, a *grind*, under common circumstances, never will break though the boats.

In the second half of the 20th century it was first understood that the stones, when continually thrown into the sea, form a wall of air bubbles behind the swimming whales. This wall reflects the whales' sonar so they think that they cannot pass the air wall and in this way they are driven to the beach (Bloch 2002, interview).

**Figure 19.** *The* kast, *used in driving pilot whales to the whaling beach* (Kerins 2006).

*Fastakast* are stones about a kilogram in weight with a hole drilled through them, through which a rope is tied. These stones are sometimes painted white and have been used for over a century (Müller 1882). It has long been thought that the whales under the water see white stones better (Williamson 1945).

Today, a modern version of the *kast* has been designed and is being produced locally for sale. This new *kast* is made from a plastic tube 30 cm long, 40 mm in diameter and about 2 mm thick, weighing approximately 0.85 kg. One end of the tube is sealed and attached to a rope approximately 12 m long. The other end is filled with a 6 cm lump of lead, which is riveted vertically in the tube. Immediately above the lead-filled end is a hole, with another smaller hole on the same side near the end of the tube. The construction results in the emission of a row of air bubbles through the small top hole when the equipment is thrown overboard into the water. The new *kast* works on the same principle as the old (Bloch 2002, interview).

**The Spear**

The *hvalvákn*, or whale spear, is a double-edged steel blade approximately 40 cm long on the end of a wooden pole about 2 m long. It is a tool that is no longer used in the hunting of pilot whales. The design of the *hvalvákn* has varied over the years. It was once used from onboard the boats to aid in beaching the whales. Joensen (1976:14) describes how the *hvalvákn* was once seen as an essential whaling tool:

> On board an '8-*mannafar,*' a boat with eight oarsmen, one should use 3 *vákn*: two men, each with a *vákn*, stand in the boat's bow and one man stands in the aft. A '10-*mannafar*' could have 4 *vákn*, two on each side. On board '6-*mannafar*' and '4-*mannafar,*' there were allowed two *vákn*, one in each end. In the whaling regulations of 1955, however, it is not specified how many *vákn* or how much other equipment is allowed in a boat since nowadays there is always enough to go around. The reason for specifying a definite number of *vákn* must have been that a *hvalvákn* was a great investment that one could not easily make.

There were specific rules for how the *hvalvákn* was to be used and where the whale should be struck. The spear was used at the end of the drive to move the whales ashore. When one of the *grindaformenn* would give the order, a certain whale in the school would be chosen, and then speared in the lower back behind the fin. This was done as it was thought the wounded whale would panic forward to the shore and beach itself, and in the chaos, the others would follow. It was also thought that the whales would not leave the bloodied area of water, hence the spearing always took place as close to the shore as possible. However, what usually occurred when the whale was speared in this way was that the wounded whale would thrash about, slamming its powerful fluke and causing damage to boats and crew. It would also swim back through the school bleeding, causing the men to spear more of the whales in the shallows, where they would have to be gaffed and hauled ashore. This required the men to wade out into cold water up to their chests among the thrashing whales. This made the task of killing the whales more difficult with the *grindamenn* trying to remain upright in the water, with their hands cold and covered in blood and whale oil, which made using the knife very dangerous. Müller (1882: 11) described the use of the spear:

> The attack, called 'holding a grind to,' is conducted thus: One boat in the foremost row pulls up to the herd before it is aware of the shallowness of the water, about five or six fathoms, according to the number of the herd, and the man in the prow wounds one individual with his lance. The wounded whale rushes into the herd, and the herd frightened, suddenly strikes ahead, sometimes right upon the shore, leaving several individuals lying on the dry shore. The first row of boats pursues and distributes thrusts with the lances, right and left, by which means the water is soon mixed with blood, as the animals are very plethoric, which has the effect that, although the grind makes a turn before it is landfast, it turns into the blood again; and it has happened that a large portion of a grind has broken through the rows and got outside the whale voe, but nevertheless has allowed itself to be driven again into the blood.
>
> The second and third rows [of boats] keep outside the clear water till it is evident that the grind 'maler': that is, the animals are bewildered and seek the blood; upon which all the boats pull into the slaughter, and with the heavy lances work an awful butchering. It is of great importance to give the first thrust or stroke at the proper instant, and with necessary attention.

By this period, the use of the spear had been a point of contention among the *grindamenn* for some time; some older men saw the tool as a necessary piece of equipment, but to be used only in particular circumstances with great skill, rather than it being a tool used by everyone in any way they chose. The nature of the slaughter being a public event has meant that the skill employed by the men has always been discussed and evaluated. At the turn of the 20th century, the skill with which the *grindamenn* carried out the slaughter was being scrutinised not only in terms of the quality of the meat and blubber but also in terms of the humane killing of the animal. Sanderson translates comments made about a hunt in Føringatíðindi in 1898:

> When the slaughter is over, the tickets have come out, and the whales have been cut up and loaded—everyone goes home again with the report card showing how the conduct has been in the kill. The report card is the blubber, in which the spears, hooks and knives have printed their clear accounts. The blubber shows whether the *grind* men have acted as brave skilful fellows, or whether they have been over-eager and irresponsible (Anonymous, *Føringatíðindi*, July 7 1898, translated in Sanderson 1992:105-106)

Others saw the spear as a problem, as seen in the following account from an experienced whaler:

> Without a doubt—it was the whaling spear that caused the biggest problem. We see today, now that the spear is no longer used, there is almost never any damage... . I also think the drive is more

> humane now than it was before. It is not easy to explain, but I think that the whales are driven more calmly into shore. The idea with the spear was to stir up the school to get them ashore quickly. But more often than not this didn't work. Now they come swimming smoothly into the bay and that means that the whole process is both quicker and more effective all around (Interview with Schønning Eysturoy (1918-1996), by Hans Jákup Hermansen 1995, translated by Kate Sanderson).

The demise of the *hvalvákn* began in the 1950s when its use was first limited in the Klaksvík whaling district. The *grindamenn* here had observed the whales were quieter and showed less resistance when the spear was not used. This in turn ensured less thrashing by the whales and less damage to the boats, making it more efficient and safer (Author's fieldnotes, December 14 2001).

**The Harpoon**

The harpoon (*skutil*) is another piece of equipment that is no longer used. Although it was once included in the pilot whaling equipment it was only ever used when the *grindamenn* were unable to secure or drive the whales on to the shore and the school was about to be lost. At this stage the *grindaformenn* could give the order that the men could harpoon the whales in the open sea before they all escaped. Williamson (1945:122) notes in the early 20th century that:

> Harpoons are often taken, but their use is strictly forbidden unless and until the captain and the district officer *[syslumaður]* decide that it is impossible to put the *grind* ashore. It is then a case of 'every boat for himself' and the harpoon is the only weapon that will capture a whale successfully at sea.

When I asked several older *grindamenn* about the use of the harpoon, they said that a pilot whale taken with a harpoon was no good as its use made the meat tough.

**The Whale Hooks**

The *sóknarongul* (whale hook) is a heavy iron hook approximately 45 cm in length and attached to a long rope that is struck into the blubber layer of the whale so that the animal can be hauled into shallower water. In the past it was also used to draw the whale closer to the boat from where it could be killed with a knife. Modern boats have higher gunwales than the traditional Faroese boats, making it more difficult for a person to lean over the side and secure the whale with the hook. This method of killing whales is no longer used and was outlawed in the 1980s.

Today a new blunt hook, called a *blástrarongul,* is used to secure the whales. This new gaff has a rounded tip at the end (Fig. 20), and is designed for insertion into either of the vestibular air sacs of the blowhole without wounding the whale (as the *sóknarongul* does), after which the whale can be easily guided in the desired direction. The advantage of the new hook is that those whales that must be secured and hauled ashore are not wounded prior to the actual killing.

The *blástrarongul* was developed by one of the pilot whalers who was anxious to improve the hunt. A new Executive Order (number 46) was implemented in 1998 that recognised the *blástrarongul* as official whaling equipment that can be used alongside the *sóknarongul*. Currently it is estimated that there is about 50 percent usage of the *blástrarongul* by pilot whalers (personal communication Sjúrðaberg 2007). The *Grindamannafelagið* (Faroese Pilot Whalers Association) ensures that each whaling district has access to the *blástrarongul*.

**Figure 20.** *The new blunt-hook (*blástrarongul*) used to haul the whale into shallow water* (Kerins 2006).

**Whaling Knives**

Pilot whales are killed with knives called *grindaknívar* which were probably the hunt's most distinguished piece of equipment (Fig. 21). Many of the knives were hand-made and highly decorated, with the sheath often inlaid with brass and Mother of Pearl shell with motifs depicting pilot whales and pilot whaling equipment.

The knife is worn about the waist and is often hung on a woven woollen band of red, white and blue. Today a *grindaknívar* is not as recognisable as in the past, for many *grindamenn* prefer imported Swedish or German designs, which have handles made of strong plastic at the base of which are small thumbs about 2 cm long to act as a safety device by stopping the hand from slipping down over the blade, an accident that can occur very easily when the knife and hand are made slippery during slaughtering or butchering a whale.

**Figure 21.** *A traditional Faroese whaling knife (*grindaknívar*), still in use today* (Kerins 2002).

The knife is used to quickly sever the blood supply to the brain. Unlike land mammals, the blood supply to the brain of pilot whales is through major arteries which enter the cranial cavity together with the spinal cord through the *foramen magnum*. For this reason a dorsal cut is made in the neck about a hand's breadth behind the blow hole (*see* Fig. 12). This is made deep enough to sever the spinal cord and the surrounding blood vessels carrying the blood to the brain, at which point the animal is paralysed and loses consciousness within 5–10 seconds (*Grindaboð* 1996:3).

**The Merkingarknívar**

The *merkingarknívar* is a small tool (Fig. 22) used for cutting the Roman (sequential) and Arabic (skinn value) numbers into the heads and flippers of pilot whales once they have been assessed. These small tools are similar to a lino-cutter and ensure the number is easily cut into the carcass.

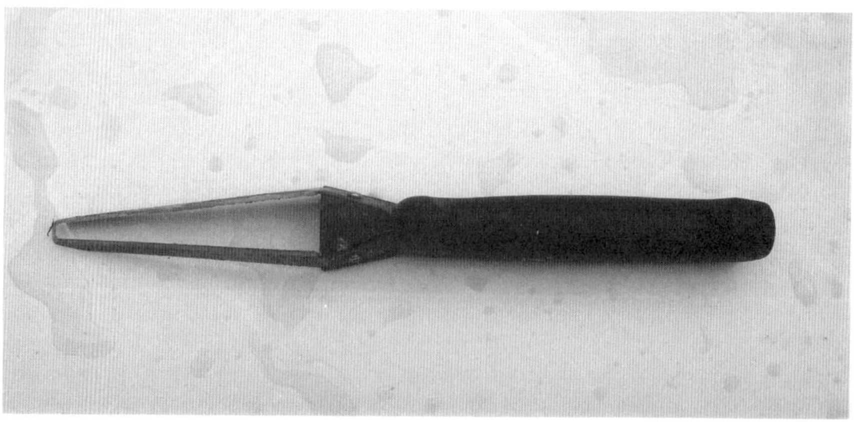

**Figure 22.** *The* merkingarknívar, *used to number the whales* (Kerins 2006).

**The Assessing Rod**

Prior to 1832 the *skinn* value of a whale was assessed visually. And, as with any visual and imprecise assessment, this often led to disagreements (Bloch and Zachariassen 1989). Because of the size of whales it was not possible to weigh the carcass, so an alternative method needed to be found. The method that was developed was a measuring rod logarithmically divided in such a way that the measured length of the whale was converted into the weight of the consumable parts of the carcass (Fig. 23). A farmer developed the measuring rod with help from a priest in 1822 in response to the dissatisfaction occurring when using a purely visual assessment (Bloch and Zachariassen 1989; Bloch 2006):

> All 40 rods have been measured and compared with the lengths and weights of pilot whales of both sexes and all reproductive stages. It is clear that the farmer and the priest were sharp mathematicians, but they also had a large amount of data at their disposal since nearly two-thirds of all landed pilot whales before 1832 belonged to the church as tithe and to the farmers as the landowner's share. Both men were also living in the village with the most often used whaling bay (Bloch 2006:65).

**Figure 23.** *A detail of the logarithmic assessing rod showing the* skinn *values* (Kerins 2006).

The original rod still exists on *Vágoy* with all later rods made using the original as the template. The measuring rod was included in the Executive Orders concerning pilot whales in 1832 and has been used since that time. The assessing rods used today are distributed in the various whaling districts and placed in the authorised whaling bays, and when one is worn out a new one is made to replace it (Fig. 23).

Faroese whaling is not an *ad hoc* 'free for all,' but a highly regulated process that is closely monitored from the sighting of whales through to the slaughter, assessment and distribution of the meat and blubber. The institution of the *grindadráp* derives acceptance and legitimacy through being embedded in state law and enjoys wide community support because the institution belongs to the Faroese people and reflects their long-held customs and practices in the hunting of whales and the distribution of the products of the hunt. Furthermore, the *grindadráp* remains economically important as it provides access to subsistence foods and enables households to be less reliant on costly imported meat.

When the *grindadráp* was threatened by the fisheries boycott launched by environmental and animal rights organisations, the Faroese people did not move away from the institution, nor did the Faroese Government seek to undermine the institution and close it down as a relic from a Viking past. What occurred was a review of the institution of the *grindadráp,* carefully examining its various components and associated issues such as the driving, beaching, slaughter, technology and wider ecological and environmental issues, after which improvements were made.

Therefore, there is clear evidence that seven of the design principles identified by Ostrom (1990:90) can be found in the *grindadráp*. Thus, we observe clearly defined boundaries where individuals have clear rights to withdraw resource units from the common property resource, where there is congruence between appropriation and provision rules and local conditions, where local people have the right to adjust operational rules, where the hunt is subject to strong and effective monitoring—with sanctions imposed when infractions occur and an adequate conflict resolution process is in place—and where the rights of resource users are respected by the state.

In the past, the Faroese could have been described as ‘passive users’ of the pilot whale resource, where their interaction with pilot whales began at the sighting of a school and ended with the consumption of the meat and blubber. Today, the Faroese are best described as ‘active managers’ of the pilot whale resource, where, the Faroese work both domestically and regionally on all management matters.

# *Nested Enterprises—Local and Regional Organisations*

---

*We believe that for marine resources such as fish and whales, and no matter what the size of the whale—international cooperation is much more likely to work on a regional level rather than a global level. Before you can really work together on a problem, you first have to be sure that you are all talking about the same thing. In NAMMCO we know that we are all talking about the same whales and seals as resources which can be used, as long as this is sustainable... . Our experience from the International Whaling Commission is that the member countries there are no longer talking about the same thing—but some are still pretending to. When we talk about conservation we do not mean that we should turn the oceans into a whale sanctuary—but we are talking about using our resources in a sustainable manner.* Jørgen Niclasen, Faroese Minister of Fisheries (WCW 1999:11).

The Faroese further strengthened the *grindadráp* by establishing a national resource-users' association to represent and advocate on behalf of Faroese whalers, and by assuming a leadership role in forming a regional marine mammal management organisation in which they could nest the local-level institution—the *grindadráp*. Nested enterprises, the eighth of Ostrom's design principles, allow local-level institutions to retain their benefits and address their limitations by nesting these institutions within a larger institutional environment that facilitates co-ordination among them, and better addresses large-scale problems that local-level organisations share in common or find costly to manage independently (Ostrom and Shlager 1996:147).

In the previous chapter, the Faroese are described as 'active managers' of whale resources, rather than as passive users as they were in the past. This is because the Faroese now work with neighbouring countries through an international regional agreement on the management of marine mammals in the North Atlantic. Under the *North Atlantic Marine Mammal Commission* (NAMMCO) *Agreement*, the Faroes, Greenland, Iceland and Norway cooperate on management-relevant matters, including the biology, ecology, population identification, social structure, and abundance, as well as hunting methods and environmental issues that affect marine mammals in the North Atlantic.

These two developments undertaken by the Faroese are important, and as Wallerstein's (1991) world-systems analysis recognises, despite variations within the core and the peripheral zones, the possibility exists that societies on the periphery of the world economy can, to some degree, through local decision-making, shape their own future in a positive manner.

## The Faroese Pilot Whalers Association—*Grindamannafelagið*

Prior to the early 1990s there was no national association in the Faroes that represented or asserted the collective rights of Faroese pilot whalers. It is not that the organisation of people into specific interest groups was a foreign idea, for the Faroese have a long history of labour unions, such as *Framtiðin* (Apostle *et al.* 2002), but in the case of pilot whaling, there was no need as no-one in the Faroes is a pilot whaler by profession. Although there has been a long tradition of holding public meetings in specific districts to discuss pilot whaling issues, there has never been a formal process to link the dispersed whaling communities into a single nation-wide organisation.

However, public meetings did play an important role in activities of the common-property resource institution, as the *grindamenn* of a particular whaling district met to elect the *grindaforemenn* whose role it was to manage the whale hunt in that particular district. The *grindamenn* also used these public meetings to discuss any other issues they may have in regard to whaling; for example issues associated with the distribution of the meat and blubber. In past years, the Minister of Fisheries and Marine Affairs also used these district meetings to communicate with the *grindamenn* of the various whaling districts concerning proposed amendments to the Executive Orders. The international protests, to some degree, acted as the catalyst for the formalisation of these meetings by linking the resource users in dispersed whaling districts together into a single representative organisation.

A formal resource-users' association would also meet another emerging need identified by Faroese whalers. That need was to also serve as a forum to pass whaling knowledge among users and between generations. For example, from the very earliest days of whaling in the Faroes, knowledge about whales, their environment, and whaling were passed on as tradition through practice. Until the mid-19th century there were no public schools, and "the learning method was watching and copying or using the phrase 'take one, be one, make one,' meaning see what a person is doing and then do it yourself and in the end get another person to take after you" (Olsen 2006:42). Even after schooling was introduced, this informal system of learning continued until the mid-20th century when young and old living in small communities spent much time together in daily activities. However, as Olsen *(ibid)* notes, in the old isolated villages in the Faroes, this parochial method of learning could also be disadvantageous, because in districts where there were few *grind*, sometimes for a period of 50 to 60 years, more than a generation of local residents would likely not participate in a *grindadráp*. Such an outcome was also partially due to people not travelling far from their home community in the days prior to the building of roads and the introduction of cars. Today, although there are roads, undersea tunnels and many cars, whaling knowledge can still be lost. When compared with the past, parents and children today do not spend as large amount of time with each other. Children are at school and their parents often work away from the villages in the capital area where other industries are situated (*op. cit*). Such a situation encouraged alternative ways of passing on practical knowledge, and, to some degree, this need could be met through establishing a representative resource-users' association. In regard to taking that action, it is reported that:

> There were a lot of heated meetings in the early 1980s, not because there was a lot of pressure coming from the outside, but because there was a lot of discussion going on internally. So when all these [animal rights] campaigns started producing all this photographic material etc., there had already been a discussion going on here to change the regulations. One of the main issues was the outlawing of the spear, which a lot of people were pushing for. You already had a committee set up at the time looking at the humane killing aspect of the hunt, because the animal welfare legislation [1984] had been adopted and it was pointed out that we had to bring the pilot whaling regulations into line with the legislation. The question was: what was the quickest and most efficient way to kill a pilot whale? So that process was going on when the real onslaught began with the protests. But I am sure there was a lot of resistance to change in some areas, so I guess a lot of these meetings were very heated and a lot of these whaling communities saw the government as coming in and interfering with their traditional way of doing things and as an affront to their skills or whatever.
>
> Then when they agreed the formation of a whalers' association was worth pursuing, we developed some draft statutes and called a meeting with representatives from all over to formally establish the association. The main idea behind it was to ensure the momentum we built up in terms of describing the hunt to the outside world and keeping a dialogue on improvements would continue in the absence of having someone deal with it on a daily basis. The thinking was that if everyone pulled together and everyone discussed the issues very seriously, then that was something that had a huge value in public relation terms outside of the Faroes. For example, we are organised; we do care what happens; we are concerned enough to review the regulations when they are not working and to discuss better ways of doing things (Sanderson 2002, interview).

The establishment of a resource-users' association was seen by the Faroese as a proactive way to influence debate and further strengthen their common property resource institution. A senior government official, in an interview, spoke about the importance of keeping the dialogue open with many of the anti-whaling organisations in order to explain the Faroese perspective and demonstrate that the Faroese had nothing to hide, saying "Whaling is legal in the Faroes and is highly managed through legislation that reflects us" (Hoydal 2006, interview).

After discussions took place between whalers and government officials, a decision was made to establish a resource-users' association, and on June 19 1993, the *Grindamannafelagið* (The Faroese Pilot Whalers Association) was officially founded in Tórshavn. Representatives from all authorised whaling bays were present at that meeting and gave their unanimous support to the creation of the association; the *Grindamannafelagið* would formalise the exchange of information between whaling districts and would aim to:

- maintain and enhance pilot whaling in order to ensure a sustainable catch of pilot whales according to need;
- co-operate internationally with other organisations with similar objectives;
- inform about pilot whales and pilot whaling nationally and internationally; and
- support scientific projects about pilot whales and their living conditions *(Grindaboð* 1996:4).

All meetings of the *Grindamannafelagið* are open to the public and anyone is welcome to attend. In 2006, there were about 1,000 members with each member paying a membership fee of 100 Faroese kroner.[1] The *Grindamannafelagið* used the long-established whaling district meetings to communicate on a regular basis with its members. Government officials, scientists, a veterinarian, and the chair and/or deputy-chair of the *Grindamannafelagið* attend the meetings in the various whaling districts. The meetings are usually held in the local village hall, school gymnasium or suitable office space. The participants are predominantly men, and a mixture of young and old attend. In the early meetings between the Government and pilot whalers in the various districts prior to the formation of the *Grindamannafelagið,* most of the participants would have been over 50 years of age. However, as the outside protests against whaling increased, more and younger people began participating in the meetings; in regard to this revival of wider community participation in pilot whaling, a senior government official observed: "I guess we have the NGOs to thank for that" (Hoydal 2006, interview).

The local meetings of the *Grindamannafelagið* are held in the evenings after work during the winter months, a time of the year when daylight is short and most men are in the village and have time to meet. The district meetings of the *Grindamannafelagið* possess social significance, for they also serve as a venue for a wide array of useful social and sometimes business interactions. Here we begin to see collective choice arrangements in operation, as well as the role cultural differences can play in the function of common property rights regimes:

> Because these meetings are organised by the local communities and they involve the various biologists, people from the government, etc. it gives it a level of credibility, rather than a government person coming along and telling the people this is how it is going to be. It's more in tune with how things have really been done over the years when change has come from the community itself rather than down from the government, from the top down. Pilot whaling has always been on a community level (Sanderson 2002, interview).

[1] Approximately US$17 in 2006.

Matters discussed at the meetings over recent years include briefings by government officials about international developments which may impact on Faroese whaling, and new whaling tools, such as the new knife developed in 2000. Scientists also take the opportunity to brief the whalers on new research projects and seek feedback or ask for whalers' cooperation in data collection and activities such as satellite tagging. Issues, such as contaminant loading in marine mammals and the health implications of consuming the meat and blubber are increasingly on the agenda of the meetings. In addition to the district meetings, which the *Grindamannafelagið* hold during the winter months, an annual general meeting is held in Tórshavn, usually in the spring. This provides the opportunity for many of the *grindamenn* from across the whaling districts, as well as supporters of the *Grindamannafelagið*, to receive reports on the association's past year's activities and finances.

The *Grindamannafelagið* have also been active developing educational material on Faroese whaling practices. In 2005, the *Grindamannafelagið* released a DVD entitled *Whaling in the Faroes*, which portrays the sighting, driving, slaughtering and distribution of the meat and blubber. Most importantly, the DVD contains interviews with older, experienced hunters who demonstrate flensing techniques and discuss local knowledge about whaling and the technology used and how to handle the meat and blubber to avoid wastage. The DVD also details processing, recipes, and techniques for safely storing the meat and blubber.

Representatives of the *Grindamannafelagið* also attend regional and international meetings to exchange experiences and information, and to help inform other whaling communities about the role of pilot whales in the daily life and livelihoods of the Faroes (*Grindaboð* 1996). An example of this role can be seen where representatives of the *Grindamannafelagið* were active participants in the formation and development in 1998 of the global organization, the World Council of Whalers (WCW 1998a), with representatives of the *Grindamannafelagið* participating in successive general assemblies of the WCW, culminating in the Faroese Government, in partnership with the *Grindamannafelagið*, hosting the fourth General Assembly of the WCW in Tórshavn in 2002 (WCW 2002). Although discussion of the WCW is outside of the scope of this book, an outline of its development and Faroese participation at its general assemblies is provided in Appendix 1.

## North Atlantic Marine Mammal Commission

In responding to a perception of uncertainty about whale population estimates in the late 1980s and early 1990s, the Faroese Government invested heavily in the formation of a regional marine-mammal management organisation that would gather and analyse data to assist with decision-making at both national and regional levels. This organisation was the North Atlantic Marine Mammal Commission (NAMMCO), and membership in this commission provided the Faroese with the opportunity to share management costs associated with issues such as large-scale sighting surveys used to collect data on whale species' distribution and abundance estimates.

NAMMCO, as the name suggests, is the authoritative regional organisation in the North Atlantic for cooperation on the conservation, management and study of marine mammals. It provides a mechanism for cooperation on conservation and management for all species of cetaceans (whales, porpoises and dolphins) and pinnipeds (seals and walruses) in the North Atlantic region. Many of these species have never been subject to any regional intergovernmental agreement; polar bears are the only marine mammal that NAMMCO does not manage within the North Atlantic region, as this species has a separate international agreement.[2] An overview of the genesis of NAMMCO, its legal basis and organisational structure, is provided in Appendix 2 (*see also* Hovelsrud-Broda 2005).

NAMMCO, with its membership drawn solely from North Atlantic nations, is viewed by the region's whaling communities in Greenland, the Faroes, Iceland and Norway as being highly informed about local 'northern issues' (Andresen 1997). It is seen as open to change and not afraid to examine new marine management ideas based on expert scientific advice and users' knowledge, underpinned by a policy of sustainable use and a multi-species approach to management. With anti-whaling groups excluded, regional whaling communities assert that management decisions do not become stalemated by conflict over opposing positions on sustainable use of marine mammals (Bendiksen 2001, interview).

The inaugural meeting of NAMMCO was held in Tórshavn in 1992, with much international attention focused on the meeting as the establishment of such an intergovernmental organisation was perceived by some states as an attempt to weaken the IWC. Demonstrating its intention to base management decisions on sound scientific data, the initial meeting was devoted to the organisation of the Scientific Committee. It was agreed that each contracting party was to have three members on the Scientific Committee; the principal task of this 12-person committee was to provide advice to the Council based on the best available scientific findings.

Through regional cooperation, the member countries of NAMMCO aimed to strengthen and further develop effective conservation and management measures for marine mammals. It was agreed that such measures, while based on the best available scientific information, should also take into account both the complexity and vulnerability of the marine ecosystem and the rights and needs of coastal communities to make a sustainable living from what the sea provides (Sanderson 1997). One of the first tasks undertaken by NAMMCO was to develop a list of priority species to be reviewed by its Scientific Committee. Pilot whales were included on the list and the Scientific Committee began a comprehensive review of the species, examining the status and abundance of pilot whale stocks as well as distributional, genetic and morphometric (body measurement) evidence (NAMMCO 1997). In addition, examinations of a

---

[2] The *International Agreement on Conservation of Polar Bears and Their Habitat* was ratified in 1976 by the five range states of polar bears: Canada, Denmark (for Greenland), Norway, the U.S. and the former-U.S.S.R.

number of environmental issues, such as contaminant loading, were also undertaken (NAMMCO 2000).

Faroese whalers and the Faroese Government took seriously some of the issues raised by the animal rights and environmental organisations, such as, issues associated with humane killing methods and sustainability of the catch. The NAMMCO Council, at its second meeting in 1993, specifically requested the Scientific Committee to analyse the effects of Faroese whaling on North Atlantic pilot whales, especially whether the numbers taken were consistent with sustainable utilisation (NAMMCO 1997). After a review of the available data, NAMMCO concluded that an annual catch of 2,000 pilot whales represented an exploitation rate of 0.26 percent of the best estimate of the population abundance of pilot whales in the northeast Atlantic (778,000 pilot whales) and was therefore sustainable (NAMMCO 1998a).

At the 1998 NAMMCO meeting, the Faroes informed the Council that revised dietary recommendations for the consumption of pilot whale meat and blubber were recently issued by health authorities in the Faroes, based on the demonstrated effects of mercury exposure and a general assessment of polychlorinated biphenyls (PCBs). The meat of the pilot whales is high in mercury and is considered to be one of the main sources of mercury in the Faroese diet, while the blubber contains high levels of PCBs. Research on the toxicity and effects of PCBs in the diet is still being undertaken. The revised regulations were a matter of considerable concern to the Faroese for whom pilot whale meat and blubber remain an important part of the household diet and which possesses important nutritional value (NAMMCO 1999b). The issue of contaminant loading in marine mammals, especially whales, is emerging as a major threat to the institution of the *grindadráp* and the free provision of nutritious fresh local food of high cultural and symbolic importance (Kalland *et al.* 2005:122).

In conclusion, the institution of the *grindadráp* was strengthened by the Faroese in two particular ways: first, by establishing a national resource-users' association, the *Grindamannafelagið* in order to help maintain, enhance and transmit traditional whaling knowledge between generations in the Faroes and to represent and advocate, nationally and internationally, on behalf of Faroese whalers. Second, by establishing, through regional cooperation, an international marine-mammal management organisation (NAMMCO), in which the local-level institution (the *grindadráp*) could nest without loosing any of its assigned rights or responsibilities.

These initiatives were taken in response to both internal and external concerns. Internally, the Faroese needed, *inter alia,* to safeguard traditional whaling knowledge, and to further strengthen the collective choice arrangements and gain legitimacy, particularly in the international arena, for their local-level common property resource institution. In regard to external concerns, the Faroese needed to address some of the criticism of their whaling practices made by animal rights and environmental organisations in their international protests, as well as nullifying attempts by these organisations to remove local decision-making over small cetaceans by placing decision-making within a global, highly-politicised, organisation (the IWC) far removed from and insensitive to the resource users' circumstances and needs.

The Faroese Government achieved these goals through taking an active role in establishing and supporting the development of a cooperative regional science-based marine mammal governance arrangement, the North Atlantic Marine Mammal Commission (NAMMCO), and by responding to external criticism of Faroese whaling, particularly in regard to pilot whale population abundance and the nature and conduct of the hunt.

# Conclusion

---

> *Conscious of being a small people living at the periphery of the inhabitable world, the Faroese find that many of the protests from people of metropolitan areas reflect not only inadequate information on the subject, but also a regrettable lack of understanding of Faroese conditions, and a lack of respect of the right of the Faroese people to utilise their natural resources.* (Olafsson 1988).

The purpose of this study has been to examine the Faroese common property resource institution, the *grindadráp*, within a broad political, economic, social and ecological framework in order to assess the *grindadráp* in terms of its ability to adapt and endure over time and to sustainably manage the resource on which it is based (the North Atlantic long-finned pilot whale).

In their attacks on Faroese pilot whaling, animal rights and environmental organisations focussed on a single component of the institution of the *grindadráp*—the slaughter—and used it as a symbol of the *grindadráp*, in isolation from its other necessary social, biological and economic components, whilst portraying it negatively (by way of bloody images) as the total representation of the *grindadráp*. These protest organisations also characterised Faroese society as being antiquated, unchanging, and isolated. The historical evidence shows that Faroese society has been one that is "*always* in flux, in a perpetual historically sensitive state of resistance and accommodation to broader processes of influence that are as much inside as outside of the local context" (Marcus and Fisher 1999:78). Indeed, the animal rights and environmental organisations are themselves part of this flux, insofar as they represent one of the external influences requiring a response from Faroese society.

The research methods used in this study, combined with a strong theoretical framework of political economy and its related world-systems theory and property rights, has allowed me to demonstrate that the Faroes is not a discrete isolate with its own antiquated and outdated resource-use practices disconnected and far removed from the influences of global political and economic forces. The historical evidence indicates that the Faroes is a place where the 'outside forces' in fact are an integral part of the construction and constitution of the 'inside,' the cultural unit itself, and must be registered" (*ibid*: 77).

Among the criticisms levelled against the *grindadráp* by those opposed to this whale hunt, are claims that (1) the hunt was unsustainable, (2) Faroese history and culture provide little more than justification for unsustainable whaling, (3) a subsistence and modern economy cannot co-exist in the same society, (4) there is no economic need to hunt whales, and (5) pilot whaling was a sport, undertaken for fun. The Faroese responded to some of these criticisms, especially those directed to questions about sustainability and humane killing. In the 1980s, at the beginning of the anti-whaling campaigns, the *grindadráp* removed an average of 2,063 pilot whales each year (Desportes *et al.* 1994). Although, in relation to immediately previous years this level of take was high, in relation to numbers of whales taken extending back to the year 1584, annual removals in the 1980s compare favourably with removals in other periods of high removal rates, which appear to peak approximately every 120 years (Joensen and Zachariassen 1982). With this historical perspective in mind, the Faroese were not overly concerned about the charge their whaling was unsustainable.

However, in order to further respond to this persistent criticism of unsustainable whaling, the Faroese authorities continued to invest in scientific research on North Atlantic pilot whales, in two particular ways. The first was by engaging in an international research program on pilot whales that focussed on stock identity, reproduction, social structure, age-related parameters and the presence of parasites and pollutants in the whales. The second was through assuming a leading role in the establishment of an intergovernmental body—the North Atlantic Marine Mammal Commission (NAMMCO)—and tasking the NAMMCO Scientific Committee to undertake a comprehensive review of North Atlantic long-finned pilot whales. This review would, *inter alia,* examine the effect of Faroese whaling on the North Atlantic pilot whale population, and to determine whether the Faroese hunt was biologically sustainable. NAMMCO concluded that an annual catch of 2,000 whales represented an exploitation rate of 0.26 percent of the present best estimate of abundance of pilot whales (numbering 778,000) in the northeast Atlantic, and that the hunt was therefore sustainable (NAMMCO 1998a).

The assertion that 'whaling is a sport and little more than fun' was then addressed. Although to an outsider, the *grindadráp* may appear similar to a large sporting event, it is a well-organised, regulated, and monitored hunt in which participants have designated roles and responsibilities directed to the goal of providing food for the community. As such, slaughtering is a necessary part of this activity, and the evidence indicates that the Faroese have invested heavily in research on humane killing methods with a view to seeking and adopting improvements. Thus, during the mid-1990s, the Faroese Government monitored the performance of the hunt and developed a new blunt hook, used in the pre-slaughter phase of the hunt that does not injure the whale. Research in this area included trials of a newly-designed knife used in the slaughtering process, and studies that confirmed that the centuries-old traditional method of slaughter remains the quickest, safest, and most effective way to humanely dispatch the whales (*Grindaboð* 1996).

The statement that 'Faroese history and culture provide little more than justification for unsustainable whaling' is also unsupported by the evidence that, on the contrary, indicates the *grindadráp* has been shaped by Faroese history and culture, and the wider political economy of the Faroese relationship with Europe, extending from the original settlement of the islands by the Norse in the 9$^{th}$ century through to the present day.

The belief that 'subsistence and a modern economy cannot co-exist in the same society' is contrary to the findings of numerous anthropological studies of foraging societies around the world (*see* e.g., Peterson and Matsuyama 1991;

Freeman 1993). The fundamental characteristics of a subsistence economy, as described by Wolfe (1983), are evident within Faroese society, and as elsewhere in the modern world, subsistence co-exists with the capitalist market economy. In the Faroes, whaling is nested within a wider seasonal round of production activities that are tied to the seasonal arrival and availability of marine resources. The evidence clearly demonstrates that in the Faroes today, these subsistence activities are intricately linked with the market economy in one society, where a modern capitalist commercial fishing industry rests comfortably with traditional systems of resource use which provide substantial non-commercial networks for sharing, distributing and exchanging food and other resources. The Faroese actively work to maintain this mixed economy with mutually supportive market and subsistence sectors.

The statement that 'there is no economic need to whale' is demonstrably false. Whale meat and blubber account for 30 percent of all meat produced on the Faroes (*Løgmansskrivstovan* 2006). Whale meat and blubber are important economic goods in this land of poor soils, rocky and mountainous terrain, and sometimes extreme weather, where, as a consequence, agricultural production is extremely limited. Further, whale meat and blubber is a locally-produced food staple which does not require transporting over long distances—a significant environmental consideration in relation to today's concern for reducing global carbon emissions. Access to this inexpensive local food staple affords Faroese households a significant degree of food security, especially during times of economic uncertainty.

The assertion that 'the use of modern technology threatens sustainability of the pilot whales' is also false. Motorboats were first used in the driving and beaching of whales at the beginning of the 20$^{th}$ century and provided the opportunity to increase the extent of whaling operations. However, this increase in whaling did not occur, because Faroese whalers understand that driving pilot whales for more than four nautical miles makes the whales more intractable and, consequently, the hunt less successful. The shared view of the Faroese, in regard to the most efficient way to hunt pilot whales, ensures that new technologies are assessed in relation to the knowledge and traditions that inform the 1,000-year-old institution of the *grindadráp*. If one looks at technology, as Freeman (1990) suggests, as artifice, rather than artefact, we see the knowledge that people use for practicable purposes as having more significance than just the tools used—whether they are traditional or modern. Here, the animal rights' and environmental organisations have "inadvertently overlooked the artifice behind the technology in favour of the artefacts that it produces" (Ridington 1982:471), and that "technology should be seen as a system of knowledge rather than an inventory of objects" (*op. cit*).

The allegations made by animal rights' and environmental organisations in their opposition to Faroese whaling had two aims. The first was to raise funds to finance their activities, and the second was to stop Faroese whaling by organising an international boycott of Faroese fish and fish products. While it is not publicly known how successful these anti-Faroese campaigns were in raising funds to finance their ongoing activities, it is known that they did not succeed in stopping Faroese whaling through their international boycotts. Instead of bringing Faroese whaling to an end, quite the opposite effect occurred. The results of these external protests resulted in critical reflection and re-evaluation of the *grindadráp* by the Faroese, and the decision not to cast aside this ancient institution, but rather, to continue to develop and refine it through science-based studies and the further inclusion of the resource users within the decision-making processes. These outside pressures raised the profile of the *grindadráp* within Faroese society, which, in turn, resulted in more younger people becoming involved in the hunt and participating in the activities of the Faroese Pilot Whalers' Association.

The re-examination of the *grindadráp,* by the Faroese, also contributed to the formation of a new regional marine mammal organisation, NAMMCO, whose role it was to gather data and provide expert scientific advice to the Faroese authorities. Collection of scientific data on population abundance of the widely-distributed long-finned pilot whale for example, was not previously available and is far too costly an undertaking for a small nation to gather alone. Most importantly, the new regional organisation did not usurp the local decision-making of the Faroese communities, and by nesting the *grindadráp* within a regional framework, the new arrangements added value to the local institution through the provision of scientific and technical advice. The new regional marine-mammal commission, in addition to collecting and analysing data, also provided a forum for the exchange of information between hunters, scientists and policy advisers on improving hunting methods (NAMMCO 1999a). Such an arrangement is in sharp contrast to the IWC meetings which few whalers attend, and only rarely participate in discussions. The new regional agreement also provided a transparent international observer scheme where the *grindadráp* can be monitored by international observers.

This book reports the results of an analysis of the *grindadráp* through a historical, political, cultural, social and environmental framework, and while that analytical framework is important for understanding how the institution developed through history and how it operates in practice, it does not directly inform us of its institutional performance, nor its efficiency as an instrument for sustainable resource management.

Consequently the notion of 'design principles' is also employed during analysis of the data. Design principles are those elements or conditions that have been identified as being critically important in helping to assess the success of institutions in sustaining common property resources and securing compliance of resource users to the institutional rules and standards (Ostrom 1990:68-69). Seven of Ostrom's design principles are evident in the performance of the *grindadráp,* and the eighth design principle, relating to nested enterprises, is evident in the Faroese role in NAMMCO. When a majority of these design principles exist within a given common-property institution, that institution can be considered robust and in a strong position to maintain itself and to sustainably manage the targeted resource over time. The findings of the NAMMCO Scientific Committee, concerning the sustainability of Faroese pilot whaling, further supports the conclusion that the *grindadráp* is a robust institution that is able to manage and sustainably utilise pilot whales.

The *grindadráp* provides resource users with significant incentives, through the provision of locally produced meat and blubber, so that they continue to commit themselves to conforming to its operational rules, to monitor each other's compliance, and continue to reproduce the common property institution across generational boundaries. Indeed, the *grindadráp* has survived for more than 1,000 years, in part because its operational rules are well adapted to local conditions and are embedded in state property law which further validates and legitimates its place in Faroese society and culture. These adaptations have been driven by the local resource users who, over time, have worked to make the *grindadráp* a more efficient, enduring, resilient and equitable institution, one where the users themselves are provided the opportunity and responsibility to manage their own resources, define their needs, goals, and aspirations, and make decisions affecting their collective well-being. The Faroese exhibit a high level of capability, responsibility and accountability in the management of the *grindadráp,* an institution that is evolutionary, participatory, locale-specific, and accommodates the technical, ecological, socio-economic and environmental circumstances of this geographically-isolated maritime community.

However, despite concluding that the *grindadráp* is a robust, enduring, efficient and resilient common property resource institution, challenges to its survival remain. The first major challenge is the issue of contaminant loading in marine mammals, and the potential health implications to both the pilot whales and to those who consume their meat and blubber. The second major challenge is a renewed threat of more anti-whaling campaigns launched from postdomestic industrialised societies where people far removed, both physically, culturally and psychologically from the animals that produce the food and fibre they depend on, use their political and financial influence to seek to impose their particular views and way of life on less populous and politically weaker domestic societies.

With this in mind, it is important to remember that the Faroes is a small nation that has fiercely maintained its community-based local decision-making over its food resources. Its actions clearly demonstrate that a small nation on the periphery of the world's centres of political and economic power can shape its own future in a positive manner. The Faroese people have continued to maintain their right to determine what is relevant and meaningful to themselves, and have not allowed this right to be usurped by distant interest groups lacking knowledge and understanding of the lives and contemporary circumstances of this unique people, and without insight into the features peculiar to that people's unique history and culture

# *References*


Act of Faroese Parliament (1984). *Protection of Animals Act 1984.* Tórshavn, Faroe Islands.

Act of Faroese Parliament (1986). *Amendments to the* Whaling Act 1986. Tórshavn, Faroe Islands.

Agnew, N.M. and S.W. Pyke (1994). *The Science Game: An Introduction to Research in the Social Sciences.* 6th edition. Englewood Cliffs, N.J: Prentice Hall.

Aguilar, A., L. Jover and A. Borrell (1993). Heterogeneities in organochlorine profiles of Faroese long-finned pilot whales: indication of segregation between pods? *Report of the International Whaling Commission* Special Issue 14: 359-367.

Akimichi, T., R. Caulfield, N. Doubleday, M. Freeman, A. Kalland, B. Moeran, G. Palsson, H. Stefansson and J. Takahashi (1992). Similarities and diversity in coastal whaling operations. A comparison of small-type whaling activities in Greenland, Iceland Japan and Norway. IWC Report IWC/44/SEST 4. International Whaling Commission, Cambridge, U.K.

Alcorn, J.B. and V.M. Toledo (1998). 'Resilient resource management in Mexico's forest ecosystems: the contribution of property rights,' pp. 216-249 in Berkes, F. and C. Fölke (eds.), *Linking Social and Ecological Systems: Management Practices and Social Mechanisms for Building Resilience.* Cambridge, U.K: Cambridge University Press.

Amos, B., D. Bloch, G. Desportes, T.M.O. Majerus, D.R. Bancroft, J.A. Barrett and G.A. Dover (1993). A review of molecular evidence relating to social organisation and breeding system in the long-finned pilot whale. *Report of the International Whaling Commission* Special Issue 14: 209-218.

Andresen, S. (1989). Science and politics in the international management of whales. *Marine Policy* 13(2): 99-117.

Andresen, S. (1997). 'NAMMCO, IWC and the Nordic countries,' pp. 75-89 in Pétursdóttir, G. (ed.), *Whaling in the North Atlantic.* Reykjavík, Iceland: Fisheries Research Institute, University of Iceland.

Apostle, R., D. Holm, G. Hovgaard, Ó.W. Høgnesen, B. Mortensen (2002). *The Restructuration (sic) of the Faroese Economy: The Significance of the Inner Periphery.* Tórshavn, Faroe Islands: Samfundslitteratur Press.

Arge, S.V. and N. Hartmann (1989). The burial site of við Kirkjugarð in the Village of Sandur, Sandoy. *Froðskaparrit* 38–39: 5-21.

Arge, S.V. (1997). Í Uppistovubeitinum, site and settlement. *Fróðskaparrit* 38–39: 27-44.

Aron, W. (1988). The commons revisited: thoughts on marine mammal management. *Coastal Management* 16: 99-110.

Aron, W., W. Burke and M. Freeman (1999). Flouting the convention. *Atlantic Monthly* 283(5): 22-29.

Assies, W.J. and A.J. Hoekema, eds. (1994) *Indigenous People's Experience with Self Government.* Proceedings of the Seminar on Arrangements for Self-determination by Indigenous Peoples within Nation States, 10–11

February 1994, Law Faculty, University of Amsterdam. IWGA Document No 76. Copenhagen.

Atlantic Margin Group (1997). *The Faroe Islands: An Environmental Statement.* A Report produced by the Environment and Resource Technology Ltd. for the Atlantic Margin Group.

Balbuena, J.A., F.J. Aznar, M. Fernández and J.A. Raga (1995). 'Parasites as indicators of social structure and stock identity of marine mammals,' pp. 133-129 in Blix, A.S., L. Walløe and Ø. Ulltang (eds.), *Whales, Seals, Fish and Man.* Amsterdam, The Netherlands: Elsevier Science.

Berkes, F. (1996). 'Social systems, ecological systems, and property rights,' pp. 87-107 in Hanna, S., C. Fölke and K.-G. Mäler (eds.), *Rights to Nature: Ecological, Economic, Cultural and Political Principles of Institutions for the Environment.* Washington, DC: Island Press.

Berkes, F., ed. (1989). *Common Property Resources; Ecology and Community-based Sustainable Development.* London: Belhaven Press.

Berkes, F., D. Feeny, B.J. McCay, and J.M. Acheson (1989). The benefits of the commons. *Nature* 340: 91-93.

Berkes, F. and C. Fölke (1998). 'Linking social and ecological systems for resilience and sustainability,' pp. 1-25 in Berkes, F. and C. Fölke (eds.), *Linking Social and Ecological System: Management Practices and Social Mechanisms for Building Resilience.* Cambridge, U.K: Cambridge University Press.

Bernard, H.R. (1988). *Research Methods in Cultural Anthropology.* Thousand Oaks, CA: Sage Publications.

Birnie, P. (1985). *International Regulation of Whaling: from Conservation of Whaling to Conservation of Whales and Regulation of Whale-Watching.* New York, NY: Oceana Publications.

Bjørk, E.A. (1963). *Førøsk Bygderet III (Strandretten)* Matrikulstovan, Tórshavn, 'Hvalretten,' 161-180; Ch. XI, "Grinefangst" 181-273.

Bloch, D. (1994). *Pilot whales in the North Atlantic. Age, growth and social structure in Faroese grinds of the long-finned pilot whale, Globicephala melas.* Ph.D Thesis, Lund University Sweden.

Bloch, D. (2006). 'Scientific use of user's knowledge in Faroese pilot whaling,' in Hovelsrud, G.K. and C. Winsnes (eds.), *Conference on User Knowledge and Scientific Knowledge in Management and Decision Making, Reykjavík, Iceland, 4-7 January 2003.* Tromsø, Norway: NAMMCO.

Bloch, D., G. Desportes, K. Hoydal and. P. Jean (1990). Pilot whaling in the Faroe Islands, July 1986–July 1988. *North Atlantic Studies* 2(1-2): 36-44.

Bloch, D., G. Desportes, R. Mouritsen, S. Skaaning and E. Stefansson (1993). An introduction to studies of the ecology and status of the long-finned pilot whale (*Globicephala melas)* off the Faroe Islands, 1986-1988. *Report of the International Whaling Commission* Special Issue 14: 1-32.

Bloch, D. and L. Lastein (1993). Morphometric segregation of long-finned pilot whales in eastern and western North Atlantic. *Ophelia* 38(1): 55-68.

Bloch, D. and M. Zachariassen (1989). The 'skinn' values of pilot whales in the Faroe Islands: an evaluation and corrective proposal. *North Atlantic Studies* 1(1): 38-56

Bromley, D.W. (1989). *Economic Interests and Institutions: the Conceptual Foundations of Public Policy*. New York, NY: Basil Blackwell.

Bromley, D.W. (1991). *Environment and Economy: Property Rights and Public Policy*. Oxford, UK: Basil Blackwell.

Bromley, D.W. (1992). 'The commons, property, and common-property regimes,' pp. 3-15 in Bromley, D.W. (ed.): *Making the Commons Work: Theory, Practice, and Policy*. San Francisco, CA: Institute for Contemporary Studies Press.

Bulliet, R.W. (2005). *Hunters, Herders, and Hamburgers: The Past and Future Relationship of Human-Animal Relationships*. New York, NY: Columbia University Press.

Burke, W.T. (2001a). Politics trumps science: the international regulation of whaling. Public lecture given at the Bevan Series on Sustainable Fisheries at the School of Aquatic and Fishery Sciences, University of Washington, Seattle WA.

Burke, W.T. (2001b). 'A new whaling agreement and international law,' pp. 51-79 in Friedheim, R.L. (ed.), *Toward a Sustainable Whaling Regime*. Seattle, WA: University of Washington Press, and Edmonton, AB: Canadian Circumpolar Institute Press.

Caurant, F., C. Amiard-Triquet and J.C. Amiard (1993). Factors influencing the accumulation of metals in pilot whales (*Globicephala melas*) the Faroe Islands. *Report of the International Whaling Commission* Special Issue 14: 369-390.

Cox, S.J.B. (1985). No tragedy on the commons. *Environmental Ethics* 7: 49-61.

Dahl, S. (1971). The Norse settlement of the Faroe Islands. *Medieval Archaeology* 14: 60-73.

Dasgupta, P. (1982). *The Control of Resources*. Cambridge, MA: Harvard University Press.

Dasgupta, P. and K.-G. Mäler (1992). *The Economics of Transnational Commons*. Oxford, UK: Clarendon Press.

Davis, A. (1984). 'Property rights and access management in the small-boat fishery: a case study from southwest Nova Scotia,' pp. 133-163 in Lamson, C. and A.J. Hanson (eds.), *Atlantic Fisheries and Coastal Communities: Fisheries Decision-Making Case Studies*. Halifax, NS: Dalhousie Ocean Studies Program.

Debes, L. (1673). *Færoæ et Færoa reserata [etc.]* Copenhagen, Denmark: Mattia Jørgensen. Republished 1963, with an introduction by E. Josnsen. Tórshavn, Faroe Islands: Einars Prent og Forlag.

Debes, L. (1676). *Færoæ et Færoa reserata: That is a description of the Islands and inhabitants of Færoæ,* translated by John Sterpin, printed by E.L. for William Iles, Flower-de-Luce in Little Britain, St. Bartholmews Gate.

Debes, H.J. (1989). Problems Concerning the Earliest Settlement in the Faroe Islands. *Fróðskaparrit* 38–39: 23-34.

Degn, A. (1929). *Oversigt over fiskeriet og monopolhandelen paa Færøerne 1709-1856. Tórshavn: Varðin.*

Department of Fisheries (undated). *Whales and Whaling in the Faroe Islands*. Tórshavn, Faroe Islands: Department of Fisheries.

Department of Foreign Affairs (2003). *Whales and Whaling in the Faroe Islands.* Tórshavn, Faroe Islands: Department of Foreign Affairs, Prime Minister's Office www.whaling.fo
Desportes, G., L.W. Andersen, P.E. Aspholm, D. Bloch, and R. Mouritsen (1992). A note about a male-only pilot whale school observed in the Faroe Islands. *Fróðskaparrit* 40: 27-33.
Desportes, G., D. Bloch, L.W. Andersen and R. Mouritsen (1994). The international research program on the ecology and status of the long-finned pilot whales off the Faroe Islands. *Fróðskaparrit* 40: 9-29.
Desportes, G. and R. Mouritsen (1993). Preliminary results on the diet of long-finned pilot whales off the Faroe Islands. *Report of the International Whaling Commission* Special Issue 14: 305-324.
Djurhuss, N., ed. (1976). *J. Chr. Svabo, Indberetninger fra en reise i Færøe 1781 og 1782.* Selskbet til udgivelse af færøske kildeskrifter og studier. Copenhagen, Denmark: C.A. Reitzels Boghandel.
DOC—Department of Conservation (2006). *The Conservation of Whales in the 21st Century.* Department of Conservation, Wellington, New Zealand. www.doc.govt.nz/Conservation/001~Plants-and-Animals/003~Marine-Mammals/Whales/100~Conservation-of-whales-in-the-21st-century/100~NZ-policy-on-whales.asp
Doubleday, N.C. (1989). Aboriginal subsistence whaling: the right of Inuit to hunt whales and implications for international environmental law. *Denver Journal of International Law and Policy* 17(2): 373-393.
Doubleday, N.C. (1994). 'Arctic whales: sustaining indigenous peoples and conserving Arctic resources,' pp. 241-261 in Freeman, M.M.R. and U.P Kreuter (eds.), *Elephants and Whales: Resources for Whom?* Basel, Switzerland: Gordon and Breach.
EBCD—European Bureau for Conservation and Development (1992). *Wise Use of Marine Resources: Why Exclude Whales?* Brussels, Belgium: EBCD.
Eggertsson, T. (1990). 'The economics of control and the cost of property rights,' pp. 157-175 in Hanna, S., C. Fölke and K-G. Mäler (eds.), *Rights to Nature: Ecological, Economic, Cultural and Political Principles on Institutions for the Environment.* Washington DC: Island Press.
EIA—Environmental Investigation Agency (1991). *The Global War Against Small Cetaceans.* 2nd report. London, UK: Environmental Investigation Agency.
EIA—Environmental Investigation Agency (1994). *Under Fire.* London, UK: Environmental Investigation Agency.
EIA—Environmental Investigation Agency (1995). *Spring News.* London, UK: Environmental Investigation Agency.
EIA—Environmental Investigation Agency (1996). *The Continuing Global War Against Small Cetaceans.* 3rd report. London, UK: Environmental Investigation Agency.
EIA—Environmental Investigation Agency (2006). *Species in Peril: Cetaceans Campaign.* London, UK: Environmental Investigation Agency. http://www.eia-international.org/campaigns/species/cetaceans/what_can_i_do/

Ensminger, J. (1996). 'Culture and property rights,' pp. 179-203 in Hanna, S., C. Fölke and K.-G. Mäler (eds.), *Rights to Nature: Ecological, Economic, Cultural and Political Principles on Institutions for the Environment.* Washington DC: Island Press.

Faroe Islands Solidarity Committee (1996). *General Correspondence.* Letter to Supporters.

Fenny, D., F. Berkes, B.J. McCay and J.M. Acheson (1990). The tragedy of the commons: twenty-two years later. *Human Ecology* 18(1): 1-19.

Fontaine, P.H. (1998). *Whales of the North Atlantic: Biology and Ecology.* Montréal, QC: Editions MultiMondes.

Freeman, M.M.R. (1990). 'The significance of animals in the life of northern foraging peoples and its relevance today,' pp. 5-12 in Okada, H. (ed.), *International Symposium on Human-Animal Relationship in the North.* Abashiri, Hokkaido: Hokkaido Museum of Northern Peoples.

Freeman, M.M.R. (1993). The International Whaling Commission, small-type whaling, and coming to terms with subsistence. *Human Organisation* 52(3): 243-251.

Freeman, M.M.R. (1994). 'Science and trans-science in the whaling debate,' pp. 143-158 in Freeman, M.M.R. and U.P. Kreuter (eds.), *Elephants and Whales: Resources for Whom?* Basel, Switzerland: Gordon and Breach.

Freeman, M.M.R. (1997). Issues affecting subsistence security in Arctic societies. *Arctic Anthropology* 34(1): 7-17.

Freeman, M.M.R. (2001). 'Is money the root of the problem? Cultural conflicts in the IWC,' pp. 122-145 in Friedheim, R.L. (ed.), *Toward a Sustainable Whaling Regime.* Seattle, WA: University of Washington Press and Edmonton, AB: Canadian Circumpolar Institute Press.

Freeman, M.M.R. (2005). 'World Council of Whalers,' p. 2195 in Nuttall, M. (ed.), *Encyclopaedia of the Arctic,* Volume 3. New York, NY and London, UK: Routledge.

Freeman, M.M.R., L. Bogoslovskaya, R.A. Caulfield, I. Egede, I.I. Krupnik and M.G. Stevenson (1998). *Inuit, Whaling and Sustainability.* Walnut Creek CA and London, UK: AltaMira Press.

Freire, P. (1982). 'Creating alternative research methods: learning to do it by doing it,' pp. 29-37 in Hall, B., A. Gillette and R. Tandon (eds.), *Creating Knowledge: A Monopoly? Participatory Research in Development.* New Delhi, India: Network Series.

Friedheim, R.L. (2001). 'Introduction: the IWC as a contested regime,' pp. 3-48 in Friedheim, R.L. (ed.), *Toward a Sustainable Whaling Regime.* Seattle, WA: University of Washington Press and Edmonton, AB: Canadian Circumpolar Institute Press.

Gambell, R. (1993). International management of whales and whaling: a historical review of the regulation of commercial and Aboriginal Subsistence Whaling. *Arctic* 46(2): 97-107.

Gambell, R. (1997). 'Recent developments in the IWC Aboriginal Subsistence Whaling category,' pp. 47-66 in Pétursdóttir, G. (ed.), *Whaling in the North Atlantic.* Reykjavík, Iceland: Fisheries Research Institute, University of Iceland.

Gannon, D.P., A.J. Read, J.E. Craddock, K.M. Fristrup and J.R. Nicolas (1997a). Feeding ecology of the long-finned pilot whale in the western North Atlantic. *Marine Ecology Progress Series* 148: 1-10.
Gannon, D.P., A.J. Read, J.E. Craddock and J.G. Mead (1997b). Stomach contents of long-finned pilot whales (*Globicephala melas*) stranded on the U.S. mid-Atlantic coast. *Marine Mammal Science* 13: 405- 418.
Gibbs, C.J.N. and D.W. Bromley (1989). 'Institutional arrangements for management of rural resources: common property regimes,' pp. 22-32 in Berkes, F. (ed.), *Common Property Resources: Ecology and Community-based sustainable development.* London, UK: Belhaven Press.
Gibson-Lonsdale, J.J. (1990). Pilot whaling in the Faroe Islands: its history and present significance. *Mammal Review* 20(1): 44-52.
*Grágás* (1852). *Islædernes Lovbog i Fristatens Tid*, vol. I udgivet efter det kongelige Bibliotheks Haandskrit, og oversat af Vilhálmur Finsen for det nordiske Literatur-Samford. Copenhagen, Denmark: Brødrene Berlings Bogtrykkeri.
*Grindaboð* (1996). *The Newsletter of Whales and Whaling in the Faroe Islands,* Number 2. Tórshavn, Faroe Islands: Department of Fisheries, Government of the Faroe Islands.
*Hagstova Føroya* (2000). *Statistical Bulletin 2000 of the Faroe Islands.* Hagstova Føroya, Argir, Faroe Islands.
Hallenstvedt, E. and G. Blichfeldt, eds. (1995). *Additional Essays on Whales and Man.* Reine, Norway: High North Alliance.
Hamilton, J.R.C. (1964). 'Brochs and broch builders,' pp. 53-90 in Wainwright, F.T. (ed.), *The Northern Isles.* Edinburgh, Scotland: Thomas Nelson.
Hammond, P.H. (1993). Letter of resignation as Chairman of the Scientific Committee of the IWC, 26 May 1993 www.highnorth.no/Library/Management_Regimes/IWC/le-fr-th.htm
Hanna, S. (1999). 'Property rights and marine ecosystems,' pp. 116-122 in Schei, P.J., O.T. Sandlund and R. Strand (eds.), *Proceedings: The Norway/UN Conference on the Ecosystem Approach for Sustainable Use of Biological Diversity.* 6-10 September 1999, Trondheim, Norway. Trondheim, Norway: Directorate for Nature Research and Norwegian Institute for Nature Research.
Hanna, S., C. Fölke and K.-G. Mäler (1999). 'Property rights and the natural environment,' pp. 1-12 in Hanna, S., C. Fölke and K-G. Mäler (eds.), *Rights to Nature: Ecological, Economic, Cultural and Political Principles on Institutions for the Environment.* Washington, DC: Island Press.
Hansen, B. (2002). 'The sea,' pp. 112-133 in Kruse, B. (ed.), *Hunters of the North: Fangskultur i Vestnorden.* Tórshavn, Faroe Islands: Forlagið Sprotin.
Hansen, S.S. (2003). 'The early settlement of the Faroe Islands: the creation of cultural identity,' pp. 73-111 in Barrett, J.H. (ed.), *Contact, Continuity, And Collapse: The Norse Colonisation of the North Atlantic.* Studies in the Early Middle Ages, Volume 5. Centre for Medieval Studies, University of York, UK.

Happynook, K., ed. (2004). *Whaling Around the World.* Brentwood Bay, BC: World Council of Whalers.
Hardin, G. (1968). The tragedy of the commons. *Science* 162: 1243-1248.
Headland, T.N. and L.A. Reid (1989). Hunter-gatherers and their neighbors from prehistory to the present. *Current Anthropology* 30: 43-66.
Heazle, M. (2006). *Scientific Uncertainty and the Politics of Whaling.* Seattle, WA: University of Washington Press and Edmonton AB: Canadian Circumpolar Institute Press.
High North Alliance (1994). *High North News* No. 9. Reine, Norway.
High North Alliance (2000). The right to be commercial. *The International Harpoon* No. 3. Reine, Norway.
Hoel, A.H. (1990). Norwegian marine policy and the International Whaling Commission. *North Atlantic Studies* 2(1–2): 117-123.
Hoel, A.H. (1993). Regionalisation of international whale management: the case of the North Atlantic Marine Mammal Commission. *Arctic* 46(2): 116-123.
Holm, D. and B. Mortensen (2002). 'Economic life on the periphery. A comparative study of economic development in two towns on Suðuroy,' pp. 27-73 in Apostle, R., D. Holm, G. Hovgaard, Ó.W. Høgnesen and B. Mortensen (eds.), *The Restructuration (sic) of the Faroese Economy: The Significance of the Inner Periphery.* Tórshavn, Faroe Islands: Samfundslitteratur Press.
Holt, S. 1985. Whale mining, whale saving. *Marine Policy* 9(3): 192-213.
Hovelsrud-Broda, G. (2005). 'North Atlantic Marine Mammal Commission,' pp. 1456-1457 in Nuttall, M. (ed.), *Encyclopedia of the Arctic.* New York, NY and London, UK: Routledge.
Hovgaard, G. and R. Apostle (2002). 'Case study of socio-economic change: analytical concerns,' pp. 13-22 in Apostle, R., D. Holm, G. Hovgaard, Ó. W. Høgnesen and B. Mortensen (eds.), *The Restructuration (sic) of the Faroese Economy: The Significance of the Inner Periphery.* Tórshavn, Faroe Islands: Samfundslitteratur Press.
Hoydal, K. (1995). Letter from Kjartan Hoydal, Director of Fisheries, to Mr. Glenross of the Independent Television Commission. *Føroya Landsstýri,* Tórshavn, Faroe Islands.
Hoydal, K. and L. Lastein (1993). Analysis of Faroese catches of pilot whales (1709-1992), in relation to environmental variations. *Report of the International Whaling Commission* Special Issue 14: 89-106.
Høgnesen, Ó.W, and R. Apostle (2002). 'Economic alternatives in an industrial Faroese town,' pp. 77-100 in Apostle, R., D. Holm, G. Hovgaard, Ó.W. Høgnesen and B. Mortensen (eds.), *The Restructuration (sic) of the Faroese Economy; The Significance of the Inner Periphery.* Tórshavn, Faroe Islands: Samfundslitteratur Press.
ICRW—International Convention for the Regulation of Whaling (1946). Washington, DC. http://www.iwcoffice.org/commission/convention.htm
ICES—International Council for the Exploration of the Sea (1996). *Report of the Study Group on Long-Finned Pilot Whales.* ICES 1996/A6.
ICES—International Council for the Exploration of the Sea (2005). About us—ICES Convention. http://www.ices.dk/indexnofla.asp

ICES—International Council for the Exploration of the Sea (2006). Our Members. http://www.ices.dk/aboutus/ourmembers.asp

*International Fieldsports and Conservation* (1991). When the sea turns red. July–August, 1991: 45-47.

International Study Group on Norwegian Small-Type Whaling (1992). *Norwegian Small Type Whaling in Cultural Perspectives*. Tromsø, Norway: Norwegian College of Fisheries.

IWC—International Whaling Commission (1982). *Report of the Special Working Group of the Technical Committee Concerning Management Principles and Development of Guidelines about whaling for Subsistence by Aborigines,* IWC Document IWC/40/5. Cambridge, UK: International Whaling Commission.

IWC—International Whaling Commission (1993). Circular Communication to Commissioners and Contracting Governments. G/VJH/19770, 18 March, Annex A. Cambridge, UK: International Whaling Commission.

IWC—International Whaling Commission (2001). Proceedings of the 51st Meeting of the IWC. Cambridge, UK: International Whaling Commission.

IWC—International Whaling Commission (2004). IWC Information History and Purpose. http://www.iwcoffice.org/commission/iwcmain.htm#history .

IWC—International Whaling Commission (2007). Population Estimates. http://www.iwcoffice.org/conservation/estimate.htm

Jakobsen, J. (1904). *Færøsk Sagnhistorie*, H.N. Tórshavn, Faroe Islands: Jacobsens Forlag.

Jakobsen, J., ed. (1907). *Diplomatarium Færoense: Føroyskt fodnbrævasavn. I: Miðalaldarbrøv upp til trúbótarskeið, við søguligun rannsókunn.* Tórshavn, Faroe Islands and Copenhagen, Denmark: H.N. Jacobsen Bókahandil and Vilhelm Prior.

Joensen, J.P. (1976). Pilot whaling in the Faroe Islands. *Ethnologia Scandinavica* 1976: 1–42.

Joensen, J.P. (1988). The pilot whale in the old and modern society. Paper for the conference “Man and the Animal World,” 22-23 September 1988, Nijmegen, The Netherlands.

Joensen, J.P. (1990). Faroese pilot whaling in the light of social and cultural history. *North Atlantic Studies* 2(1–2): 179-184.

Joensen, J.S., and Zachariassen P. (1982). Pilot whale statistic 1584-1640 and 1709-1978. *Report of the International Whaling Commission* 40: SC/42/011.

Johnsen, A.O. (1962). Hvalfangst: Norge. *Kulturhistorisk Leksikon for Nordisk Middelalder* Vol. III, pp. 160-164. Copenhagen, Denmark: Rosenkilde og Bagger.

Johnston, G., translator (1994). *Thrand of Gotu: Two Icelandic Sagas from the Flat Island Book.* Erin, ON: Porcupine’s Quill.

Kalland, A. (1993). Management by totemization: whale symbolism and the anti- whaling campaign. *Arctic* 46(2): 124-133.

Kalland, A. (1994). ‘Aboriginal subsistence whaling: a concept in the service of imperialism,’ pp. 5-11 Blichfeldt, G. (ed.), *11 Essays on Whales and Man.* 2nd edition. Reine, Norway: High North Alliance.

Kalland, A. (2010). Fangst av grindehval I Norge–vs. Færøyene (Pilot whaling in Norway vs. the Faroe Islands). *Fróðskaparrit* 58. bók 2010:100-124

Kalland, A., F. Sejersen, H.B. Broch and M. Ris (2005). *Marine Mammals and Northern Cultures*. Edmonton, AB: Canadian Circumpolar Institute Press.

*Kommisionsbetænkningen* 1709-10 (1934). *Færoernes Tilstand ved Kongens Overtagelse af Enchandelen paa Færoene.* Tórshavn, Faroe Islands.

*Kunngerð um Grind* (2000). *Havlir og Grind.* Føroya Landsstýri, Tórshavn, Faroe Islands.

Kuper, A. (1999). *Culture: An Anthropologist's Account.* Cambridge, MA: Harvard University Press.

Landt, J. (1965). *Forsøg til en beskrivelse over Færøerne 1800.* Tórshavn, Faroe Islands: Einars Prent og Forlag.

Langdon, S.J., ed. (1986). *Contemporary Alaskan Native Economies*. Lanham, MD: University Press of America.

Lawrence-Neuman, W. (1994). *Social Research Methods. Qualitative and Quantitative Approaches*. Boston, MA: Allyn and Bacon.

Lynge, F. (1992). *Arctic Wars, Animal Rights, Endangered Peoples*. Hanover, NH and London, UK: University Press of New England.

*Løgmansskrivstovan* (2006). *The Faroe Islands. A North Atlantic Perspective on Sustainable Development.*, Prime Minister's Office, Tórshavn, Faroe Islands.

Macpherson, C.B. (1978). 'The meaning of property,' pp. 1-13 in Macpherson C.B. (ed.), *Property: Mainstream and Critical Positions.* Toronto, ON: University of Toronto Press.

Marcus, G.E and M.J. Fisher (1999). *Anthropology as Cultural Critique: An Experimental Movement in the Human Sciences.* Chicago, IL and London, UK: University of Chicago Press.

Marine Protection Group (undated). Newsletter. UK.

Martin, A.R. and P. Rothery (1993). Reproductive parameters of female long-finned pilot whales (*Globicephala melas*) around the Faroe Islands. *Report of the International Whaling Commission* Special Issue 14: 263-304.

Martínez Cobo, J.R. (1983). *Study of the Problem of Discrimination against Indigenous Populations.* U.N. Economic and Social Council, Human Rights Commission. U.N. Doc. E/CN.4/Sub.2/1983/21/Add.8. New York, NY: United Nations.

Mawer, G.A. (1999). *Ahab's Trade: The Saga of South Seas Whaling.* Sydney, NSW: Allen and Unwin.

McCay, B.J., and J.M Acheson, eds. (1987). *The Question of the Commons: The Culture and Ecology of Communal Resources.* Tucson, AZ: University of Arizona Press.

McCay, B.J. (1996). 'Common and private concerns,' pp. 111-126 in Hanna, S., C. Fölke, K.-G. Mäler (eds.), *Rights to Nature: Ecological, Economic, Cultural and Political Principles on Institutions for the Environment.* Washington D.C: Island Press.

McLay, J.K. (1995). *The International Whaling Commission: Wither or Whither?* Lecture to TerraMedia World of Whales Exhibition Lecture Series, 21 August 1995, Auckland, New Zealand.

McLay, J.K. (1996). *Whales: what future for the International Whaling Commission (IWC)?* Opening remarks to Consultative Seminar for Non-Governmental Organisations (NGOs), industry, political parties and media. Stafford House, The Terrace, Wellington, New Zealand.
Mercer, M.C. (1975). Modified Leslie-Delury population models of the long-finned pilot whale (*Globicephala melas*) and annual production of short-finned squid (*Illex illecebrosus*) based upon their interaction in Newfoundland. *Journal of the Fisheries Research Board of Canada* 32: 1145-1154.
Morton, H. (1982). *The Whale's Wake*. Dunedin, New Zealand: University of Otago Press.
Müller, H.C. (1882). Whale Fishing in the Faroe Islands. Prize Essays of the International Fisheries Exhibition, Edinburgh, Scotland. 1-19.
NAMMCO—North Atlantic Marine Mammal Commission (1992). The NAMMCO Agreement. Tromsø, Norway: NAMMCO. www.nammco.no/Nammco/Mainpage/DocumentsAndInformation/nammco_agreement.html
NAMMCO—North Atlantic Marine Mammal Commission (1997). *NAMMCO Annual Report 1996.* Tromsø, Norway: NAMMCO.
NAMMCO—North Atlantic Marine Mammal Commission (1998a). *NAMMCO Annual Report 1997.* Tromsø, Norway: NAMMCO.
NAMMCO—North Atlantic Marine Mammal Commission (1998b). 'Report of the fifth meeting of the Scientific Committee,' pp. 85-202 in *NAMMCO Annual Report, 1997.* Tromsø, Norway: NAMMCO.
NAMMCO—North Atlantic Marine Mammal Commission (1999a). Rules of Procedure for the Committee on Hunting Methods . Tromsø, Norway: NAMMCO http://www.nammco.no/webcronize/images/Nammco/749.pdf
NAMMCO—North Atlantic Marine Mammal Commission (1999b). *NAMMCO Annual Report 1998.* Tromsø, Norway: NAMMCO.
NAMMCO—North Atlantic Marine Mammal Commission (2000). *NAMMCO Annual Report 1999.* Tromsø, Norway: NAMMCO.
NAMMCO—North Atlantic Marine Mammal Commission (2001). *NAMMCO Annual Report 2000.* Tromsø, Norway: NAMMCO.
NAMMCO—North Atlantic Marine Mammal Commission (Undated). *Status of Marine Mammals in the North Atlantic. The Long-finned Pilot Whale.* Tromsø, Norway: NAMMCO.
Nauerby, T. (1996). 'The pilot whale hunt: from national symbol to international stigma,' pp. 144-169 in Nauerby, T. (ed.), *No Nation is an Island. Language, Culture, and National Identity in the Faroe Islands.* Aarhus, Denmark: Aarhus University Press.
Nelson, D. and J. Lien (1996). The status of the long-finned pilot whale, *Globicephala melas*, in Canada. *Canadian Field Naturalist* 110: 511-524.
Nørlund, N.E. (1944). *Færøernes Kortlægning: En historisk fremstilling.* Goedætisk Instituts Publikationer VI. Copenhagen, Denmark: Ejnar Munksgaard.
North, D.C. (1990). *Institutions, Institutional Change and Economic Performance.* New York, NY: Cambridge University Press.

Oakerson, R. (1992). 'Analyzing the commons: a framework,' pp. 41-59 in Bromley, D. (ed.), *Making the Commons Work: Theory, Practice, and Policy.* San Francisco, CA: Institute for Contemporary Studies Press.
Ohmagari, K. (2005). 'Whaling conflicts: the international debate, pp. 145-178 in Kishigami, N. and J.M. Savelle (eds.), *Indigenous Use and Management of Marine Resources.* Senri Ethnology Studies 67. Osaka, Japan: National Museum of Ethnology.
Olafsson, Á. (1988). Account of the Pilot Whale Hunt in the Faroe Islands. Paper from Adviser on Faroe Island Affairs in the Danish Foreign Ministry, Copenhagen, Denmark.
Olafsson, Á. (1990). Faroese whale and whaling policy. *North Atlantic Studies* 2(1–2): 130-137.
Olsen, J. (2006). 'Hunting activities in the Faroe Islands: how user knowledge is gathered, kept and transmitted among pilot whale hunters in the Faroe Islands,' pp. 38-43 in Hovelsrud, G.K. and C. Winsnes (eds.), *Conference on User Knowledge and Scientific Knowledge in Management and Decision Making.* Reykjavík, Iceland 4-7 January 2003. Tromsø, Norway: NAMMCO.
Ostrom, E. (1990). *Governing the Commons: The Evolution of Institutions for Collective Action.* New York, NY: Cambridge University Press.
Ostrom, E. (1992). *Crafting Institutions for Self-Governing Irrigation Systems.* San Francisco, CA: Institute for Contemporary Studies Press.
Ostrom, E. and E. Schlager (1996). 'The formation of property rights,' pp. 127-156 in S. Hanna, C. Fölke and K.-G. Mäler (eds.), *Rights to Nature: Ecological, Economic, Cultural and Political Principles on Institutions for the Environment.* Washington, DC: Island Press.
Ostrom, E., J. Walker and R. Gardner (1992). Covenants with and without a sword: self-governance is possible. *American Political Science Review* 86: 404-417.
Ostrom, E. and J. Walker (1997). 'Neither markets nor states: linking transformation processes in collective action arenas,' pp. 35-73 in Mueller, D.C. (ed.), *Perspectives on Public Choice: A Handbook.* New York, NY: Cambridge University Press.
Oxford English Dictionary (1995). Oxford, UK: Oxford University Press.
Payne, P.M. and D.W. Heinemann (1993). The distribution of pilot whales (*Globicephala spp.*) in shelf/shelf-edge and slope waters of the northeastern United States, 1978-1988. *Report of the International Whaling Commission* Special Issue 14: 51-68.
Peterson, N. and T. Matsuyama, eds. (1991). *Cash, Commoditisation and Changing Foragers.* Senri Ethnological Studies No. 30. Osaka, Japan: National Museum of Ethnology.
Pinkerton, E., ed. (1989). *Co-operative Management of Local Fisheries: New Directions for Improved Management and Community Development.* Vancouver, BC: University of British Columbia Press.
Poulsen, J.H. and U. Zachariassen, eds. (1971). *Seyðabrævið.* Tórshavn, Faroe Islands: Føroya Fróðskaparfelag.

Rasmussen, J. (1982). 'The Faeroe Islands: geology,'in Rutherford, G.K. (ed.), *The Physical Environment of the Faeroe Islands*. The Hague, The Netherlands: Dr. W. Junk Publishers.
Rice, D.W. (1998). *Marine Mammals of the World: Systematics and Distribution*. Special Publications Number 4, Society for Marine Mammalogy.
Ridington, R. (1982). Technology, world view, and adaptive strategy in a northern hunting society. *Canadian Review of Sociology and Anthropology* 19: 469-481.
Runge, C.F. (1992). 'Common Property and Collective Action in Economic Development,' in Bromley, D. (ed.), *Making the Commons Work: Theory, Practice, and Policy*. San Francisco: Institute for Contemporary Studies Press.
Rutherford, G.K., ed. (1982). *The Physical Environment of the Faeroe Islands*. The Hague, The Netherlands: Dr. W. Junk Publishers.
Sanderson, K. (1990). *Grindadráp*. The discourse of drama. *North Atlantic Studies* 2(1–2): 196-204.
Sanderson, K. (1992). *Grindadráp. A Textual History of Whaling Traditions in the Faroes to 1900*. Unpublished Master of Philosophy Thesis. University of Sydney, Sydney, Australia.
Sanderson, K. (1997). 'The North Atlantic Marine Mammal Commission – in principle and practice,' pp. 67-74 in Pétursdóttir, G. (ed.), *Whaling in the North Atlantic*. Reykjavík, Iceland: Fisheries Research Institute, University of Iceland.
Scarce, R. (1990). *Eco-Warriors: Understanding the Radical Environmental Movement*. Chicago, IL: Noble Press.
Schultz, E.A. and R.H. Lavenda (2001). *Cultural Anthropology: A Perspective on the Human Condition*. Mountain View, CA: Mayfield Publishing.
Sea Shepherd Conservation Society (2000). *Campaigns—The Faroe Islands Grindstop 2000*. www.seashepherd.org/campaigns/grindstop2000/intobreach.html
Sea Shepherd Conservation Society (2006). About Sea Shepherd Conservation Society. http://www.seashepherd.org/about-sscs.html
Sergeant, D.E. (1962). On the external characters of the blackfish or pilot whales (Genus *Globicephala*). *Journal of Mammalogy* 43: 395-413.
Sigurjónsson, J. and G.A. Víkingsson (1997). Seasonal abundance of and estimated food consumption by cetaceans in Icelandic and adjacent waters. *Journal of the Northwest Atlantic Fishery Science* 22: 271-287.
Skov, H., J. Durinck, F. Danielsen and D. Bloch (1995). Co-occurrence of cetaceans and seabirds in the North East Atlantic. *Journal of Biogeography* 22: 71-88.
Spencer, L., J. Bollwerk and R. Morais (1991). The not so peaceful world of Greenpeace. *Forbes,* November 11, 1991: 174-177.
Stewart, I. (1991). Statement by Mr I L G Stewart. New Zealand Commissioner to the International Whaling Commission and Advisor to the New Zealand Delegation. United Nations Conference on Environment and Development, Third Session of the Preparatory Committee, Geneva, 12 August - 4 September 1991. Working Group II: Item 2.

Storm, G., ed. (1881) *Samlede Skrifter af Peder Claussøn Friis*. Kristiania, Norway: A. W. Brøgger.

Svabo, J.C. (1959). *Indetetninger fra en Rejse i Færøe*. København, Denmark.

Svabo, J.C. (1779). Om den Færøske Marsviin-Fangst. *Det almindelige Danske Bilbiothek. Et Maaneds-Skrift 3*. København, Denmark.

Sørensen, H. (1994). 'The environmental movement and minke whaling,' pp. 27-30 in Blichfeldt, G. (ed.), *11 Essays on Whales and Man*. Reine. Norway: Nigh North Alliance. www.highnorth.no/Library/Movements/General/th-en-mo.htm

Tierney, J.J., ed. and translator (1967). *Liber De Mensura Orbis Terrae*. Scriptores Latini Hiberniae, 6. Dublin, Ireland: Dublin Institute for Advanced Studies.

*The Times* (1996). 1 May 1996. London, U.K.

Tod, F. (1982). *Whaling in Southern Waters*. Dunedin, New Zealand: Frank Tod.

Tønnessen, J.N. and A.O. Johnsen (1982). *The History of Modern Whaling*. London, UK: Hurst.

Tyril, B. (2006). *Annual Business Report*. The Official Report of the Faroese Business Review. Viborg, Denmark: Nørhaven Books.

UNEP/GRID-Arendal (2004). Whaling districts in the Faroe Islands, *UNEP/GRID-Arendal Maps and Graphics Library*, http://maps.grida.no/go/graphic/whaling_districts_in_the_faroe_islands

Usher, P. J. (1981). 'Sustenance or recreation? The future of native wildlife harvesting in Northern Canada,' pp. 56-71 in Freeman, M.M.R. (ed.), *Proceedings of the First International Symposium on Renewable Resources and the Economy of the North*. Ottawa, ON: Association of Canadian Universities for Northern Studies and Canada Man in the Biosphere Programme.

Vialles, N. (1994). *Animal to Edible*. Cambridge, UK: Cambridge University Press.

Wallerstein, I. (1989). *Modern World System III—The Second Era of Great Expansion of the Capitalist World Economy, 1730–1840s*. New York, NY: Academic Press.

Wallerstein, I. (1991). *Unthinking Social Science: The Limits of Nineteenth-Century Paradigms*. Cambridge, MA: Polity Press.

Watson, L. (1981). *Guide to Whales of the World*. London, UK: Hutchinson.

WCED—World Commission on Environment and Development (1987). *Our Common Future: Report of the World Commission on Environment and Development*. Oxford, UK: Oxford University Press.

WCW—World Council of Whalers (1998a). 'World Council of Whalers. Constitution and History,' pp. 79-109 in *Report on the First Annual General Assembly of the World Council of Whalers, Victoria, BC, Canada 2-4 March 1998*. Brentwood Bay, BC: WCW

WCW—World Council of Whalers (1998b). *1998 General Assembly Report, Victoria, B.C. Canada. April 2-4 1998*. Brentwood Bay, BC: WCW.

WCW—World Council of Whalers (1999). *General Assembly Report*. Reykjavík, Iceland, March 27–30, 1999. Brentwood Bay, BC: WCW.

WCW—World Council of Whalers (2000a). *Report on the World Council of Whalers' Assembly and Conference: Securing Food, Health, and Traditional Values through the Sustainable Use of Marine Resources. Nelson, New Zealand.* November 16-19, 2000. Brentwood Bay, BC: WCW.

WCW—World Council of Whalers (2000b). *Tohora* Newsletter of the World Council of Whalers, Te Ohu Kai Moana and Nga Iwi o te Tau Ihu o te Waka a Maui. Te Ohu Kai Moana. Wellington, New Zealand.

WCW (World Council of Whalers (2002). *Report of the World Council of Whalers' General Assembly and Conference: The Future of Whaling. Tórshavn, Faroe Islands,* September 26-29, 2002. Brentwood Bay, BC: WCW.

WDCS—Whale and Dolphin Conservation Society (undated). Fundraising Correspondence.

WDCS—Whale and Dolphin Conservation Society (1994). Fundraising Correspondence.

WDCS—Whale and Dolphin Conservation Society (1997). Newsletter, May.

WDCS—Whale and Dolphin Conservation Society (2001). *The Faroe Islands.* www.wdcs.org/dan/publishing.nsf/allweb/99ACA139A9913609802568C400324D74

Weinberg, A. (1976). Science in the public forum: keeping it honest. *Science* 191: 341.

Weinstein, M.S. (2000). Pieces of the puzzle: solutions for community-based fisheries management from Native Canadians, Japanese cooperatives and common property researchers. *The Georgetown International Environmental Law Review* 12(2): 375-412.

Wenzel, G. (1991). *Animal Rights, Human Rights: Ecology, Economy and Ideology in the Canadian Arctic.* Toronto, ON: University of Toronto Press.

West, J.F. (1972). *Faroes: The Emergence of a Nation.* London, UK: Hurst.

White, J.B. (1985). Law as rhetoric, rhetoric as law: the arts of cultural and communal life. *University of Chicago Law Review* 52: 684.

Williams, M. (2003). *Making Sense of Social Research.* London, UK: Sage.

Williamson, K. (1945). The economic and ethnological importance of the caaing whale, *Globioephalus melaena* Traill, in the Faroe Islands. *North Western Naturalist* 20: 118-136.

Williamson, K. (1948). *The Atlantic Islands.* London, UK: Routledge and Kegan Paul.

Wolf, E.R. (1997). *Europe and the People Without History.* Berkeley, CA: University of California Press.

Wolfe, R.J. (1983). 'Understanding resource use in Alaskan socioeconomic systems,' pp. 248-274 in Wolfe, R.J. and L.J. Ellanna (compilers), *Resource Use and Socioeconomic Systems: Case Studies of Fishing and Hunting in Alaskan Communities.* Juneau, Alaska: Alaska Department of Fish and Game, Division of Subsistence, Technical Paper No 61.

Wylie, J. (1987). *The Faroe Islands: Interpretations of History.* Lexington, KY: The University Press of Kentucky.

Wylie, J. (1995). 'The White Man's Book. The sense of time, the social construction of reality, and the foundations of nationhood in Dominica and the Faroe Islands,' pp. 31-70 in Hughes, D.O. and T.R. Trautmann (eds.), *Time: History and Ethnologies.* Ann Arbour, MI: University of Michigan Press.

Wylie, J. and D. Margolin (1981). *Ring of Dancers: Images of Faroese Culture.* Philadelphia, PA: University of Pennsylvania Press.

Yin, R.K. (1984). *Case Study Research: Design and Methods.* Thousand Oaks, CA: Sage.

Young, O.R., M.M.R. Freeman, G. Osherenko, R.R. Andersen, R.A. Caulfield, R.L. Friedheim, S.J. Langdon, M. Ris and P.J. Usher (1994). Subsistence, sustainability, and sea mammals: reconstructing the international whaling regime. *Ocean and Coastal Management* 23:117-127.

Zachariassen, P. (1993). Pilot whale catches in the Faroe Islands, 1709-1992. *Report of the International Whaling Commission* Special Issue 14: 69-88.

## Films

Environmental Investigation Agency. (1995). *Animal Detectives, Pilot Whaling In the Faroe Islands.* Produced by Goldhawk in association with Eco-detectives, United Kingdom.

Pilot Whalers' Association (2002). *Whaling in the Faroe Islands.* Produced by Videomynir for the Faroese Pilot Whalers Association, Faroe Islands.

## Interviews

Bendiksen, Bjørn Hugo. (Norwegian fisher, minke whaler)

Bloch, Dorete. Dr. (Biologist)

Schønning Eysturoy (1918-1996) interviewed by Hans Jákup Hermansen 1995, translated by Kate Sanderson).

Hermansen, Hans Jákup. (Teacher, pilot whaler)

Hoydal, Kjartan. (Past Director, Faroese Department of Fisheries)

Mikkelsen, Bjarni. (Biologist, Faroese representative to NAMMCO Scientific Committee, pilot whaler)

Olsen, Jústines Dr. (Veterinarian, Faroese representative on NAMMCO Committee on Inspection and Observation)

Sanderson, Kate. (Faroese Government Official)

Sjurðarberg, Ólavur. (School principal, Chairperson of *Grindamannafelagið*, pilot whaler)

# Appendices

## Appendix 1: *The World Council Of Whalers*

The origin of the WCW can be traced to a series of meetings, the first of which occurred during the 1992 meeting of the IWC, held in Glasgow, Scotland, where a small group of community representatives from Canada, the Faroes, Greenland, Iceland, Japan, Norway, and St Vincent and the Grenadines[1] met to discuss their common concerns regarding anti-whaling protest actions taken against their communities, and the threat such actions posed to their economic and food security and cultural identity.

At this first meeting, participants were unanimous regarding the importance of whaling to the present and future social, cultural, and economic wellbeing of their coastal communities. They were also concerned about what they perceived as the conflict between the IWC Scientific Committee's assessment (as well as those of other scientific advisory bodies), which indicated that many stocks of whales can support a sustainable harvest,[2] and IWC actions which ignore this advice by maintaining the whaling moratorium (WCW 1998a). Participants asserted that emotional controversies over whaling and the lack of understanding of the importance of whales and whaling to their communities threatened the integrity and effectiveness of the scientific research programmes which were essential to the rational management and conservation of whales. In a bid to rectify this situation, community representatives proposed that an international alliance of whaling communities be formed with the long-term objective being the continuation and/or resumption of whaling on a sustainable basis, recognising the needs of coastal communities, respecting the principles of conservation and taking account of marine ecosystem dynamics (WCW 1998a:17).

Following this initial meeting, community whaling representatives held a second meeting during the 1993 IWC meeting in Kyoto, Japan, under the title of the Second World Conference on Community-based Whaling. The aim of this meeting was to further advance the formation of a global alliance of whaling

---

[1] Taking part in this meeting were representatives from the Inuvialuit Game Council, Canada; Faroese Department of Fisheries; KNAPK [Greenlandic Hunters' and Fishermen's Association]; Kristjan Loftsson, Iceland; Mayor of Oshika, Japan; Japan Small-type Whaling Association; North Norwegian Small-type Whalers Union; Sunmore Fiskarlag, Norway; High North Alliance, Norway; Norwegian Fishermen's Association, Norway; Stuart Nanton, St Vincent and the Grenadines IWC Commissioner (WCW 1998a:18)

[2] These harvests include minke whales off the coasts of Japan and Norway; minke, fin, pilot whales off Greenland; minke, fin, and sei whales off Iceland; pilot whales in the Faroes; bowhead whales off the coast of Alaska and Chukotka (Russia); and humpback whales off the coast of St Vincent and the Grenadines (WCW 1998a:16).

communities. Here, the group was made up of representatives from Canada, the Faroes, Greenland, Japan, Norway, and St. Vincent and the Grenadines, as well as some non-government organisations and scientists. Disillusionment with the IWC was again expressed at this meeting (WCW 1998a: 20). The group again urged the IWC to recognise the political and economic contribution sustainable community-based whaling makes to habitat and resource conservation, and to take immediate action to restore the economic viability of these coastal communities by issuing science-based whaling quotas (WCW 1998a:22-28).

A third international meeting was held in 1996, in association with the 6th Global Meeting of the International Association for the Study of Common Property (IASCP) held in Berkeley, California, to further progress the implementation of a global whalers' alliance. The diversity and size of the group was larger, with 35 participants from 12 countries.[3] At this meeting, the Third World Conference on Community-based Whaling, it was agreed to form an organisation to link together the world's community-based whalers. The Nuu-Chah-Nulth First Nation from Canada agreed to provide a home for the organisation's secretariat (WCW 1998a:39; Freeman 2005).

In 1996 the World Council of Whalers (WCW) was legally incorporated in Canada as an international community-based organisation whose primary role was to provide a forum to enable the world's whaling communities to exchange information on whaling, sustainable use, killing methods, health and the marine environment, and varied cultural, social, economic, and technical issues related to whaling.

As an organisation representing community-based whalers, the content and procedures of WCW meetings are strikingly different from that of the IWC. Many of the participants come directly from whaling communities, and discuss issues with government officials and scientists, as opposed to the IWC where it is primarily government officials who participate.

The widening gap between domestic and post-domestic societies was clearly evident at WCW assemblies. The WCW Chair drew attention to this gap when he said:

> As coastal people… who variously live in areas where prospects of obtaining our daily food from the land is often severely restricted by geography and climate, the resources of the sea have very special meaning to us. Many people living in cities—where too many people scarcely understand how their food is produced—are in need of education, not just about us, but also about their own food security problems (WCW 1998b:51).

Perhaps the most salient comment on the contradictions that whaling peoples see between the domestic and postdomestic societies came from Ingmar Egede, the Greenlandic President of the Katuaq Cultural Centre, in Nuuk, when he stated:

---

[3] Canada, Greenland, Indonesia, Japan, New Zealand, Norway, Russia, Samoa, St. Vincent and the Grenadines, Tonga, U.K., U.S.A

> The world is full of double standards, where value systems clash. In the IWC, the anti-whaling countries are obsessed with the concepts of 'humane killing.' Yet, those same countries produce millions of eggs, and chickens, and hogs in ways that cause one to wonder if the animals in question ever have a decent life. These same countries export chemical wastes across international boundaries, but place all kinds of restrictions upon countries which harvest marine mammals and want to trade between nations, in marine mammal products (WCW 1998b:48).

As with the formation of the NAMMCO, Faroese representatives have fully participated in WCW General Assemblies. They were active in networking and informing people of the role of Faroese institutions in managing the hunting of whales and the distribution of their products freely to the communities, as well as strategies that had been developed to counter the animal rights organisations' boycotts of the Faroese economy. This was one of the first occasions outside of the NAMMCO framework when members of *Grindamannafelagið* began to actively participate in international discussions.

The WCW held four General Assemblies. The first, in Victoria, B.C. (Canada) in April 1998, was attended by 100 representatives from 19 countries[4] (WCW 1998b). The Second General Assembly was held in Reykjavík (Iceland), in 1999, attended by 76 representatives from 21 countries[5]; the theme of this Assembly focused on regional management of marine mammals (WCW 1999). The Third WCW General Assembly was held in 2000 in Nelson (New Zealand), hosted by The Treaty of Waitangi Fisheries Commission (Te Ohu Kai Moana) and seven Maori tribes.[6] The theme of this meeting was "Securing food, health and traditional values through the sustainable use of marine resources." The assembly attracted more than 230 participants from 23 countries[7] including elected leaders, government officials (including current and former IWC commissioners) and representatives of national and international non-governmental organisations, researchers and students (WCW 2000a). The Fourth General Assembly was held in 2002 in Tórshavn (The Faroes) where

---

[4] Antigua and Barbuda; Australia; Canada; Dominica; The Faroes; Greenland; Grenada; Japan; Iceland; Indonesia; New Zealand; Norway; The Philippines; Russia; St Kitts and Nevis; St Lucia; St Vincent and the Grenadines; Tonga; U.S.

[5] Antigua and Barbuda; Canada; Denmark; Dominica; The Faroes; Greenland; Grenada; Japan; Iceland; Indonesia; New Zealand; Norway; The Philippines; Russia; St Kitts and Nevis; St Lucia; St Vincent and the Grenadines; Tonga; U.K; U.S.; Zimbabwe.

[6] Te Runanga o Toa Rangatira; Te Atiawa Manawhenua ki te Tau Ihu; Ngati Apa ki te Waipounamu; Ngati Koata; Ngati Rarua; Te Runanga o Ngati Kuia; Ngati Tama Manawhenua ki Te Tau Ihu (WCW 2000b).

[7] Australia; Bermuda; Canada; Cook Islands; Denmark; Dominica; Faroes; Federated States of Micronesia; Fiji; Greenland; Iceland; Japan; New Zealand; Norway; Philippines; Russia; Saint Lucia; Scotland; Switzerland; Tonga; U.S.A; Zimbabwe (Zimbabwe was facing similar issues to whaling countries in regards to its community-based hunting programmes focussed on elephants).

more than 150 people attended from 11 countries[8] as well as elected leaders, government officials (including current and former IWC commissioners), representatives of national and international non-governmental organisations, researchers and students. The theme of the Assembly was "The future of whaling" (WCW 2002).

[8] Austria; Canada; Denmark; Faroes; Greenland; Iceland; Japan; New Zealand; Norway; St Lucia; U.S.

## Appendix 2: *North Atlantic Marine Mammal Commission*

The genesis of NAMMCO dates back to the 1980s, when informal discussions between North Atlantic delegations at the International Whaling Commission (IWC) focused on two perceived management problems within the IWC (Hoel 1993). First, at a practical level, no quotas, other than for aboriginal subsistence, were being granted; and second, at an institutional level, the lack of constructive dialogue within the IWC on how to move toward the IWC Scientific Committee-approved Revised Management Scheme had brought the IWC almost to a complete standstill in discharging its core responsibilities under the International Convention on the Regulation of Whaling (ICRW). Hoel noted that while disillusionment with the IWC was one concern, there was also a far more principled sovereignty issue that needed to be addressed, *viz.,* "Who was to decide on a coastal state's utilisation of the resources in its exclusive economic zone?" (*ibid*:118).

With this undercurrent of disillusionment, the evolution of NAMMCO can be charted over five meetings by a core group of North Atlantic whaling nations that included the Faroes, Iceland and Norway. Other whaling nations, Canada, Greenland, Japan, and the former Union of Soviet Socialist Republics (USSR), were on the periphery. However, although Greenland was initially peripheral, it soon moved to join the core of nations of the group that termed itself the North Atlantic Alliance (*op. cit*).

The initial meeting of this group occurred in 1988 in Iceland during the First International Conference on Management and Rational Utilisation of Marine Mammals. The agenda for the meeting focused on the concerns identified above: the present international scheme for marine mammal management and options for change, scientific research and the dissemination of information on whaling to the wider public (*op. cit*).

In 1989 the second International Conference on Management and Rational Utilisation of Marine Mammals held in Tórshavn, in the Faroes, with the same nations participating in the discussions. At this second meeting the agenda focused on multi-species management of living marine resources, including the state of knowledge relating to multi-species management, the role of marine mammals in the ecosystem, and their management. The last agenda item was important, as during these discussions, Iceland proposed the establishment of a regional organisation for co-operative management of marine mammals in the North Atlantic. This was supported by the Faroes, which also wanted a multi-species management approach for both marine mammals and fisheries within a regional organisation for the North Atlantic and the establishment of several regional management committees under it. There was universal agreement that such an organisation would not replace any existing organisation, rather it would supplement it. Government officials were then tasked with the development of a Memorandum of Understanding (MOU) to formalise such a structure between the core nations (*op. cit*).

In 1990, a third Conference on Management and Rational Utilisation of Marine Mammals was held in Tromsø, Norway. At this conference an MOU was signed by the four core North Atlantic whaling nations: the Faroes, Greenland, Iceland, and Norway. The signing of the MOU saw the establishment of the North Atlantic Committee for Co-operation on Research on Marine Mammals (NAC). The NAC MOU focused on scientific cooperation, but not management, in regard to marine mammals in the North Atlantic. The parties were of the view that the time was not yet ripe to form a management body for marine mammals, although discussions about this next step were ongoing. In this regard, NAC can be described as the mechanism that paved the way for the later NAMMCO Agreement (Sanderson 2007, personal communication).

The preamble of the NAC MOU identified the need for conservation and management of marine mammals in the North Atlantic with the aim of achieving this through a multi-species approach. The MOU also recognised the special needs of the many Inuit communities in the region represented by the Greenland Home Rule Government.

The establishment of NAC saw one representative from each country nominated to a committee that would continue to advance regional management of marine mammals in the North Atlantic. The committee was supported by the establishment of a secretariat based at the University of Tromsø in Norway. Steps toward a multi-species management approach were immediately taken, with alignment of NAC's scientific research and advice on marine mammals with the scientific advice of fisheries agencies such as the North East Atlantic Fisheries Organization (NEFCO) and the Northwest Atlantic Fisheries Commission (NWFCO) within the North Atlantic region. These linkages occurred via the development of a 'working relationship' with the International Council for the Exploration of the Seas (ICES)[9] in relation to the role of marine mammals in the ecosystem and the management of marine mammals in a multi-species context. The ICES was seen as the most appropriate body for this relationship as it was the body that provided scientific advice to member states on the management of North Atlantic fisheries. The ICES had also previously established a study group on pilot whales, and the NAC members were cautious to not unnecessarily duplicate research, but rather to build on previous findings.

In 1991, a fourth Conference on Management and Rational Utilisation of Marine Mammals was held in Reykjavík in Iceland. Here, NAC convened a parallel meeting to discuss the issue of small cetaceans. There was concern among the NAC members about the IWC's creeping jurisdiction over small cetaceans, and that they were becoming a major agenda item within the IWC with the risk of small cetacean management becoming highly politicised (Hoel 1993). The NAC member countries held the view that the IWC had no legal

[9] ICES was formed under the 'Convention for the International Council for the Exploration of the Sea' in 1964. ICES is the organization that coordinates and promotes marine research in the North Atlantic. Its membership comprises Belgium, Canada, Denmark, Estonia, Finland, France, Germany, Iceland, Ireland, Latvia, Lithuania, the Netherlands, Norway, Poland, Portugal, Russia, Spain, Sweden, the U.K. and the U.S. Each of the member countries elects two delegates who represent their country on the ICES Council. The ICES Council is the principal policy and decision-making body of ICES. A number of countries have affiliate status with ICES including: Australia, Chile, Greece, New Zealand, Peru, and South Africa (ICES 2005, 2006).

authority to manage small cetaceans under its founding charter (the ICRW), so they opposed this move by the IWC and favoured instead a regional framework for the management of small cetaceans based on a multi-species ecosystem management approach. The NAC meeting also included discussions on the inclusion of seals within the regional management institution and how management would work in practice through the establishment of several sub-regional management committees (*op. cit*).

At the 1991 IWC meeting, two events occurred that acted as the catalyst for moving NAC from a regional body with a science focus into a more robust management institution. The first of these two events was when the majority of the member nations of the IWC refused to follow the recommendation of its Scientific Committee to adopt the Revised Management Procedure (RMP), which would have provided the IWC with a safe and conservative method of calculating quotas which were scientifically robust. The second was Iceland giving notice of its intention to withdraw from the IWC, effective from June 1992 (*op. cit*). This meant that for Iceland to continue to whale it would need to demonstrate to the international community that it was whaling sustainably via a legitimate management institution. Although IWC claimed it was the only international institution with the authority to manage whaling, Iceland held the view that "whaling is permissible under general international law subject to the duty to conserve, a duty that under the United Nations Convention on the Law of the Sea (UNCLOS) Article 64 can be discharged through direct interaction with other states or though international organisations" (Burke 2001b:58). For the North Atlantic whaling countries the time was now ripe to put a supplementary agreement in place for the management of cetaceans in the North Atlantic, and it was decided to establish NAMMCO with a council as its supreme body, along with a secretariat, scientific committee and several sub-regional management committees.

At the fifth International Conference on Management and Rational Utilisation of Marine Mammals, held in Nuuk, in Greenland, the NAMMCO Agreement was signed by the Ministers of Fisheries of the Faroes, Greenland, Iceland and Norway, effective from July 1992. It was agreed that the scope of the NAMMCO Agreement include all marine mammals (except polar bears) in the North Atlantic. Under the NAMMCO Agreement the geographical scope of the organisation is delimited vaguely to 'the North Atlantic' (Article 9, NAMMCO 1992). Thus, stocks occurring in the waters of one of the parties in this region are of 'joint interest' to the organisation as a whole. It was also decided that NAMMCO should be represented at other international marine resource management bodies, such as IWC, North East Atlantic Fisheries Organization (NEFCO) and the Northwest Atlantic Fisheries Organization (NWFCO).

NAMMCO was legally institutionalised as an 'agreement' rather than a MOU to ensure that NAMMCO would qualify under the *United Nations Law of the Sea Convention* (UNCLOS), particularly Article 65, as one of several 'appropriate international organisations' for the conservation, management, and study of marine mammals. Another unique aspect of the NAMMCO agreement is that it is an agreement between the ministries of fisheries of member countries rather than their foreign ministries. This was due to the subordinate constitutional position of both the Faroes and Greenland in relation to Denmark. The Faroes and Greenland are not sovereign states and therefore under the Danish Constitution control over foreign policy resides with Denmark.

However, under Home Rule arrangements, both Greenland and the Faroes have relative freedom in relation to fisheries and their related foreign policy. For example, the Faroes and Greenland negotiate and sign their own bilateral fisheries agreements without the participation or approval of Denmark. "We see this as a normal part of the responsibilities of the coastal states' arrangement. Whoever considers themselves as the representative state of the stocks being discussed (herring, mackerel, blue whiting) participates" (Sanderson 2006, interview). With the ministries of fisheries signing the agreement, the Faroes and Greenland were able to participate as full members of NAMMCO in their own right.

The NAMMCO Agreement was also entered into without "prejudice to obligations of the Parties under other international agreements" (Article 9, NAMMCO 1992). This meant that as long as large whales were not managed by NAMMCO, no major compatibility problem was seen to exist with other relevant international organisations, such as IWC.

The structure of NAMMCO is relatively simple, consisting of a Council, Management Committees, and a Scientific Committee. The Council is the primary decision-making body of NAMMCO, made up of ministries signatory to the agreement. The Council meets annually to review advice requested from the Scientific Committee, and to coordinate recommendations for further scientific research. The Management Committees make proposals for conservation and management and recommendations for scientific research with respect to stocks of marine mammals within their mandate. The Scientific Committee provides scientific advice in response to requests from the Council, utilising the best available scientific information. The work of NAMMCO is coordinated and facilitated by the Secretariat based at the University of Tromsø in northern Norway.

NAMMCO also provides a forum for the exchange of information among member countries on other matters related to marine mammal conservation and management, such as hunting methods and environmental questions. For example, in a move designed to demonstrate transparency, soon after its establishment NAMMCO implemented the Joint Control Scheme for the Hunting of Marine Mammals, which included an international observer scheme for sealing and whaling activities in NAMMCO member countries.

NAMMCO recognises local management and actively supports this via its management approach. With a schedule of meetings and workshops held in various locations within regional member countries, there is active local participation in NAMMCO activities. Resource users (hunters) get to meet directly with scientists and other expert advisers, government ministers and their officials, and visiting guests (*see* NAMMCO 1999a, Bloch 2006).

## Appendix 3: *Glossary of Faroese Whaling Terms*

| | |
|---|---|
| *grind* | A school of pilot whales are termed a *grind* (pronounced with a short 'i' as in window) in the Faroese language. A single pilot whale is termed a *grindahvalur*. |
| *grindadráp* | the driving, slaughtering, assessment and division of pilot whales |
| *Grindaforemenn* | Whaling foremen |
| *grindahvalur* | A single pilot whale |
| *Grindamannafelagið* | The Faroese Pilot Whaler's Association |
| *Grindamenn* | Whalemen, Whalers |
| *grindabýti* | free division of the meat and blubber of pilot whales after a successful hunt |
| *Kunngerð um grind* | *Executive orders concerning pilot whaling; pilot whaling regulations* |
| *Skinn* | Dead whales are valued by *skinn* (literally 'skins'). This is the smallest reckoning of areas or values in land, according to the traditional Faroese system: 1 *mørk* = 16 *gyllin*, 1 *gyllin* = 20 *skinn*. Today, in reference to pilot whales, one *skinn* is equivalent to 38 kilos of meant and 34 kilos of blubber. |
| *Sýslumaður* | Sherriff |

## Appendix 4: *List of Figures*

## Appendix 5: *List of Tables*

## Appendix 4: *List of Acronyms*

| | |
|---|---|
| ARK | Animal Rights Kollective |
| BWU | Blue Whale Unit |
| EIA | Environmental Investigation Agency |
| HKWG | Humane Killing Working Group (of the IWC) |
| HNA | High North Alliance |
| ICES | International Council for the Exploration of the Sea |
| ICRW | International Convention for the Regulation of Whaling |
| IFAW | International Fund for Animal Welfare |
| ITC | Independent Television Commission |
| IWGIA | International Work Group for Indigenous Affairs |
| KNAPK | *Kalaallit Nunaanni Aalsartut Piniartullu Kattufiat* (Greenland's Fishers' and Hunters' Organization) |
| NAMMCO | North Atlantic Marine Mammal Commission |
| NEFCO | North East Atlantic Fisheries Organization |
| NMP | New Management Procedure |
| NWFCO | Northwest Atlantic Fisheries Commission |
| PCBs | polychlorinated biphenyls |
| RMP | Revised Management Procedure |
| RSPCA | Royal Society for the Prevention of Cruelty to Animals |
| SSCS | Sea Shepherd Conservation Society |
| UNCED | United Nations Conference on Environment and Development |
| UNCLOS | *United Nations Law of the Sea Convention* |
| WDCS | Whale and Dolphin Conservation Society |
| WSPA | World Society for the Protection of Animals |